Signals and Communication Technology

This series is devoted to fundamentals and applications of modern methods of signal processing and cutting-edge communication technologies. The main topics are information and signal theory, acoustical signal processing, image processing and multimedia systems, mobile and wireless communications, and computer and communication networks. Volumes in the series address researchers in academia and industrial R&D departments. The series is application-oriented. The level of presentation of each individual volume, however, depends on the subject and can range from practical to scientific.

Indexing: All books in "Signals and Communication Technology" are indexed by Scopus and zbMATH.

For general information about this book series, comments or suggestions, please contact Mary James at mary.james@springer.com or Ramesh Nath Premnath at ramesh.premnath@springer.com.

Deepanjali Mishra • Anh Ngoc Le
Zachary McDowell
Editors

Communication Technology and Gender Violence

Springer

Editors
Deepanjali Mishra
School of Humanities
KIIT University
Bhubaneswar, India

Anh Ngoc Le
IT Department
Swinburne University of Technology
(Vietnam) - FPT University
Hanoi, Vietnam

Zachary McDowell
Department of Communication
University of Illinois at Chicago
Chicago, IL, USA

ISSN 1860-4862 ISSN 1860-4870 (electronic)
Signals and Communication Technology
ISBN 978-3-031-45236-9 ISBN 978-3-031-45237-6 (eBook)
https://doi.org/10.1007/978-3-031-45237-6

This Springer imprint is published by the registered company Springer Nature Switzerland AG
The registered company address is: Gewerbestrasse 11, 6330 Cham, Switzerland

Paper in this product is recyclable.

Preface

The rapid acceleration of technological advancements and digital transformation marks this era. Communication technology remains an omnipresent entity, permeating every facet of human existence. We live in an epoch where innovations such as Artificial Intelligence, machine learning, robotics, and the Internet of Things (IoT) are not mere buzzwords, but tangible realities that shape our societies, economies, cultures, and identities (Schwab, 2016). This digital age is characterized by an intricate web of interconnected systems and infrastructures that continually influence and redefine our lives (Castells, 2010). Amid these techno-centric currents, a profound and nuanced understanding of technology becomes a necessity rather than a luxury.

Simultaneously, as we navigate this increasingly interconnected and digitized world, we grapple with ongoing and pressing issues around gender's social construction and interaction, playing a critical role in shaping personal identities and social interactions (Butler, 1990; Connell, 2002). Often misconstrued as a binary concept, gender remains a spectrum of identities and roles deeply entrenched in sociocultural contexts (Fausto-Sterling, 2000). Neither a static nor a homogeneous construct, gender remains a dynamic entity that shapes and is shaped by many socio-political, economic, and technological forces (Connell, 2002).

Whether it is the violence of Kant's "radical evil" or more akin to Derridian "violence" that probes the "infinitesimal difference" (Derrida, 2016, p. 234), the notion of "violence" here is to include a broad spectrum of ways in which the enaction of violence may emerge within communication technology. Consider, for example, the underrepresentation and marginalization of women in technology-related fields (Ashcraft et al., 2016), or the reinforcement of gender stereotypes in AI and machine learning models (Crawford & Paglen, 2019). While there are many gendered "radical evils" that perpetuate on the Internet, from overt misogyny to death threats, the nuanced ways in which gender violence plays out remain pervasive and incredibly problematic – often solipsized by the more overt evils. Given this backdrop, the interplay between communication technology and gender violence remains a fecund research theme that warrants exploration from many angles. The lens of communication technology serves as a potent prism to view and understand the dynamics of gender, providing insights into how gender permeates and is permeated by technological systems. Conversely, through the lens of gender, we can elucidate the often-hidden biases and disparities within technological practices and infrastructures.

The intricacies of the technology-gender relationship are manifold and complex. Many communication technologies that might be perceived as gender-neutral bear the imprint of gender norms and biases as it is created and implemented within sociocultural contexts that are inherently gendered (Wajcman, 2004). Consequently, the design and application of technology often reflect and perpetuate existing gender inequalities, even as they purport to advance progress and efficiency.

However, it is equally important to recognize that technology is not a mere vehicle of gender bias. It can also be a powerful tool for subverting traditional gender norms and catalyzing social change. The digital sphere has been heralded as a space of liberation and empowerment, where individuals can explore, negotiate, and perform their gender identities in ways that may be constrained in physical spaces. Still, it also helps shape our identity (see Turkle, 1995). Digital technologies, social media platforms, and online communities may offer novel avenues for the articulation and advocacy of gender rights and equality, such as publicity and branding within social media (see Marwick, 2013). But at what cost?

Of course, technology itself is gendered, influencing, and influenced by gender identities and roles. For instance, certain technologies have been stereotypically associated with masculinity (like computers or cars) or femininity (like domestic appliances), reflecting and reinforcing the "making" of gender norms and inequalities (Cockburn & Ormrod, 1993). At the same time, the gendering of technology is not fixed but fluid, subject to shifts and contestations in response to socio-cultural changes and technological innovations.

This re-configuring of "violence" re-orients questions we ask under this banner. How does gender influence the design, use, and impact of communication technology? How does communication technology reciprocally shape our understanding and performance of gender? How has communication technology been used to inflict or perpetrate gender violence? How can we use this knowledge to create more equitable technologies and social structures? How does gender influence the "making" of technology's design, utilization, and societal impacts (such as in Cockburn & Ormrod, 1993)? How does technology mold our comprehension and articulation of gender (like Haraway's 1991 Cyborg Feminism)? What is the significance of this knowledge in creating equitable technological systems and social structures (like the data feminism of D'Ignazio & Klein, 2020)? These and many more questions open exploration.

By offering diverse perspectives on these pivotal questions, this anthology paints a complex picture of the convergences and divergences at the nexus of technology and gender. Aimed at a diverse audience, including academics, students, technology professionals, and all those interested in the crossroads of technology and gender, this anthology invites engagement with the multidimensional analyses within these pages and extends the discussion on the essential issues they underscore.

The book begins with an overview which begins to probe the aforementioned questions. "Technology and Gender Violence: Victimization Model, Consequences and Measures" (Chap. 1) discusses the significant role of technological evolution in improving human life across various domains. However, it also highlights the dark side of technology by facilitating violence, mainly targeting women, girls, and transgender individuals through online platforms. The authors reviewed existing literature and presented a gender violence model focusing on the types of abuses, victim populations, and technologies used in these incidents as well as the consequences faced by the victims and various initiatives to prevent and manage technology-dependent gender violence. Lastly, it covers measures to sensitize and equip women and girls with the knowledge to avoid and handle such violence effectively.

Next, in "Implication of Technology to Curb Digital Violence among women on Social Media Platforms," (Chap. 2) Mishra et al. focus on gender violence, traditionally associated with violence against women but has now expanded to affect individuals of various gender identities. They discuss how technology, originally designed to be user-friendly, has become a tool for abuse and exploitation. They aim to study the impact of technology on digital violence and find ways to prevent such incidents, especially among students who are more active social media users. They highlight the need to use technology to escape the trap of abuse and lead an everyday life again.

In "Cybersecurity Analysis and Phishing Attack," (Chap. 3) Sahoo explores the growing role of the Internet of Things (IoT) in healthcare and discusses the networked healthcare and medical architecture. This chapter emphasizes the importance of safeguarding healthcare data from hackers, given its significant value as a target. Analyzing international regulations on medical and healthcare cybersecurity, it presents a normative hierarchical model of international cybersecurity standards to create a secure healthcare IoT solution. The chapter introduces a case-oriented technique, including Advanced Security Assurance Cases (ASAC), for assessing the cybersecurity of such systems. Additionally, it reports on an internal evaluation that targeted hospital staff to address phishing and healthcare, summarizing relevant literature.

Next, "Sentimental Analysis of Quarantine Fears Among Women Through NVIVO" (Chap. 4) discusses the sentiment analysis of quarantine fears experienced by working women. The study gathered qualitative data from 30 individuals working in the Education sector. The main findings of the chapter focus on how educators cope with the challenges of quarantine during Covid-19 while fulfilling their job responsibilities, which can affect their mental well-being. The analysis is aided by word cloud and sentimental analysis, along with thematic analysis using NVIVO-12. The implications highlight the importance of maintaining mental strength to deal with individual fears during quarantine in the Indian context.

Continuing the work on algorithmic biases, "Gender Biasness – A Victim of Artificial Intelligence-Based Development" (Chap. 8) discusses the concept of Algorithm Bias, Artificial Intelligence Bias, or Machine Learning Bias, which occurs when machine learning algorithms produce systematically prejudiced results due to vague assumptions. Gender bias is a common form of Artificial Intelligence bias, as cognitive biases from human developers can influence machine learning models and training data. Biases in algorithms and incomplete data lead to gender biases in the results. Ensuring that Artificial Intelligence datasets are representing the complete population and overcoming built-in biases is crucial. Currently, people's behavior and opinions are influenced, knowingly or unknowingly, by Artificial Intelligence. Gender inequality can be observed in technology design patterns due to the overrepresentation of men in the system. The chapter focuses on the role of biased algorithms in Artificial Intelligence decision-making processes, particularly in HR and higher management, as organizations rely on Artificial Intelligence to manage human resources and achieve long-term goals.

In "A Comparative Analysis of Pornography Detection Models to Prevent Gender Violence," (Chap. 9) Mishra and Panda discuss how the globalization of technology has led to a revolution in social media, bringing the world closer together and enabling easy and accessible communication. Platforms like YouTube allow users to upload videos, some of which are educative and informative, while others contain sensitive and impermissible content like pornography. Controlling the distribution of such sensitive content is crucial, especially considering the widespread accessibility of the internet. Technological innovations like Deep Learning and Artificial Intelligence have led to various pornography detection models that have been successful to some extent in restricting viewers from accessing inappropriate content. This chapter aims to study different pornography detection models and explore their role in preventing Gender Violence.

Regarding more explicit gendered violence, in "Trolls to Cyber Mob: Reasons of Trolling on Women," (Chap. 7) Karthika's research explores the phenomenon of real-life misogyny spilling into the virtual world, where men harass women through various means such as verbal abuse, obscene videos, and morphed photos. The study aims to shed light on the reasons behind the gender-based trolling of women. It establishes a connection between gender trolling, cyber violence, cyber victimization, and the psychological well-being of the participants. The research uses qualitative methods and gathers responses through three focus group discussions to examine the impact of gender trolling on women in Kerala.

On the other side of this violence, "Leveraging OSINT and Artificial Intelligence, Machine Learning to Identify and Protect Vulnerable Sections of Society" (Chap. 5) discusses potentials for protection against these issues. Despite the internet serving as a global hub for vast amounts of data storage and sharing, certain sections of society, particularly women and children, remain incredibly vulnerable to cyberbullying, harassment, and cyberstalking. To protect them from these psychological attacks, laws are being developed and modified to address cyber threats. Open-Source Intelligence (OSINT) utilizes publicly available data, such as news clippings and social media posts, which can be processed using tools to gather information. Artificial Intelligence and machine learning (AI and ML) aim to make computers more human-like by enabling them to learn human behavior. This chapter analyzes potential solutions to tackle cyber threats against women and children, proposing using Open-Source Intelligence and Artificial Intelligence technologies to detect and address cybersecurity breaches and safeguard vulnerable sections of society.

In another approach to consideration of AI and algorithmic systems, Dutta in "Framing the Landscape of Technological Enhancements: Artificial Intelligence, Gender Issues, and Ethical Dilemmas" (Chap. 10) discusses the emergence of digital technology, particularly Artificial Intelligence (AI) and machine learning (ML), and its impact on various fields. This chapter highlights the concerns related to human ethics, gender biases, social inequalities, and job opportunities arising from AI advancements. The focus is on AI's potential and implications rather than its current state. It emphasizes that AI technology aims to mimic human decision-making and behavior, and its widespread applications raise ethical and social questions. One crucial aspect explored is the gender bias present in AI development, which can perpetuate and amplify existing societal inequalities. The chapter advocates for gender-sensitive technology considering diversity, fairness, inclusivity, and individualism. It argues for the responsible and anti-discriminatory design and use of AI algorithms to ensure a humane and equitable human-computer interaction.

Approaching communication technology from a pedagogical lens, "English Lecturers' Digital Resources Use at Universities of Nepal Amidst Unsecured Online Environment" (Chap. 11) explores the use of Information Communication Technology (ICT) in tutoring literacy at Nepali universities, particularly among English speakers. The research assesses the penetration of ICT in the classroom and evaluates differences in ICT tool preferences and related tasks among preceptors despite similar preferences in traditional classroom settings. The study aims to identify the effectiveness, challenges, possibilities, pros, and cons of using ICT in English lectures. Additionally, it investigates the safety of online tutoring and literacy platforms. The research was conducted qualitatively, relying on primary data sources through virtual questionnaires sent to 30 preceptors from 12 universities in Nepal. Out of the 30 approached, 21 responses were received. The study found that using ICT in tutoring literacy is relatively common among preceptors and scholars in the classroom, with enthusiasm for adopting advanced technologies. However, it also reveals that the current practices are limited and suggests possibilities for further improvements such as dramatizing course content, role-playing, tonal studies, exploration, and distance literacy.

Addressing potentials for utilizing communication technology in emancipatory ways, Multani in "Cyberfeminism, Gender Dynamics and Women Empowerment" (Chap. 12) explores the impact of cyberspace technology on gender relations and patriarchal hegemony. It highlights how Information and Communication Technologies (ICT) empower women, facilitate feminist activism, and challenge male dominance. Cyberspace and feminism merge to create a new feminist paradigm, and various cyberfeminist discourses organize feminist, political, and cultural environments on the internet. Women's involvement in digital technologies has generated politically engaged feminist praxis, eroding patriarchal structures. However, the internet's role in the new millennium is paradoxical, acting as a liberating and constraining force for women. The chapter delves into the emancipatory potential of technology for women's empowerment and the obstacles they face. It also explores the reconfigurations of female identity through digital means, augmenting the politics of presence.

Addressing advertising and popular media representations, Sarita in "Cyber Risk and Gender Violence in Fashion Advertising" (Chap. 6) notes that the present situation in India shows that women face violence and discrimination in various sectors, with 30% of women aged 15–49 experiencing physical and mental harassment, and around 6% facing sexual violence. Despite legislative progress, gender discrimination and violence persist both publicly and in private spaces. Many cases remain hidden or underreported due to societal barriers and obstacles. To address this social evil, spreading awareness is crucial, and advertising and fashion industries can play a significant role in doing so. They have raised awareness about discrimination against women and encouraged organizations to speak up against it. International fashion forums have initiated dialogues on the harmful effects of gendered advertising on women and society. United Nations-Women Organization data shows that 35% of women worldwide have experienced physical or sexual violence from a partner or non-partner at least once in their lifetime. The media, including the fashion world, influences our perception of gender and societal roles, playing a crucial role in constructing or deconstructing gender equality.

Addressing new changes in the workplace, Huda in "Enhancing Digital Leadership Direction: Insight into Empowering Gender Violence Prevention" (Chap. 13) explores the significance of digital leadership in ensuring organizational stability, especially during the pandemic age. It emphasizes the importance of online competence skills in driving strategic coordination and facilitating flexibility and responsiveness within the organization. The focus is on developing digital professional skills to enhance leadership quality. The chapter examines the role of professional soft skills and digital competence in fostering innovation, openness, and responsiveness in managing customer relationships. The value of the chapter lies in continuously expanding digital competencies among employees while promoting trust and appreciation in the process. Overall, it highlights the essential role of digital leadership in navigating the challenges of the digital era and fostering organizational success.

Approaching the issues from an ethical standpoint, "Understanding of Digital Ethics for Information Trust: A Critical Insight into Gender Violence Anticipation" (Chap. 14) focuses on exploring the importance of digital ethics as a strategic foundation to enhance safety and foster partnerships within the digital community. It conducts a literature review, drawing insights from peer-reviewed journals, proceedings, chapters, and books related to the topic. The findings highlight the crucial role of digital ethics in promoting digital partnership skills and fostering a sense of community in the digital realm. The chapter contributes valuable insights into the effective implementation of digital ethics as a safety strategy, leading to the development of a robust digital community and successful partnerships.

This volume comprises a rich tapestry of theoretical perspectives and empirical research, including essays from diverse academic fields. The writings not only dissect the prevailing gender norms and biases embedded in current technologies but also probe the transformative potential of technology in challenging gender binaries and inequities.

Reflecting the complexity of the topic, "Communication Technology and Gender Violence" curates a wide array of theoretical lenses and empirical studies from multiple disciplines. To address these issues requires cutting across traditional academic boundaries and engaging a plethora of approaches – communication, sociology, computer science, philosophy, psychology, cultural studies, and critical feminist and queer theory. The book seeks to challenge prevalent gender norms and biases entrenched in technological systems while also uncovering the transformative potential of communication technology to challenge and reconfigure these turgid systems (see Balsamo, 1996; Turkle, 1995).

The anthology captures a diverse array of themes under its overarching narrative. It navigates through the complex gender dynamics in digital spaces, which are often unseen (see Gray, 2018), critically scrutinizes the gendered distribution of labor in technology industries, particularly in the age of work-at-home (see Hochschild, 1989 for an early approach), and assesses the representation and influence of gender in the rapidly evolving field of Artificial Intelligence (AI). By offering broad and deep insights into the multifarious issues at the intersection of technology and gender, the book contributes significantly to the burgeoning scholarship in this domain.

In acknowledging the complex realities of gender and technology, the book underscores the principle of intersectionality. It highlights that technology and gender do not exist in social vacuums. Instead, they are intrinsically intertwined with other social axes, such as race, class, age, sexuality, and disability, collectively shaping individuals' experiences and identities (Hill Collins, 2000). By engaging with an intersectional approach, the anthology offers a nuanced understanding of the social dynamics surrounding technology and gender. It allows for a more inclusive and holistic exploration of this intricate interface.

By illuminating the dynamic relationship between gender and technology, "Communication Technology and Gender Violence" aims to incite thoughtful examination and contribute to the ongoing academic and societal dialogues around gendered technology. It serves as a catalyst for further scholarly inquiry and for the development of more equitable technologies and associated policies (see Eubanks, 2018; Noble, 2018).

This book is situated at the crossroads of these technological and gendered currents. It strives to offer an exploration of the intricate relationship between technology and gender, weaving together diverse scholarly perspectives. This volume aims to provide a holistic perspective on the convergence of technology and gender by examining how these elements interact, shape, and transform each other. In doing so, we challenge established narratives and prompt critical discourse on the implications for social justice, equity, and inclusion.

The primary goal of this book is to provide a comprehensive exploration of the complex relationship between technology and gender. This exploration is two-pronged: It seeks to understand how gender structures, informs, and influences technology and, conversely, how technology reciprocally shapes, affects, and constructs gender (Wajcman, 2004). In presenting these diverse and detailed narratives, the anthology challenges existing paradigms and triggers critical reflections on the broader implications for social justice, equality, and inclusivity.

In conclusion, "Communication Technology and Gender Violence" offers a comprehensive and insightful exploration of the reciprocal relationship between technology and gender. It presents a rich tapestry of interdisciplinary scholarship that dissects, challenges, and reconstructs the traditional narratives in this field. It is the hope that this book will inspire intellectual curiosity and pragmatic action toward fostering a more equitable technological landscape.

Chicago, IL, USA Zachary McDowell

References

Ashcraft, C., McLain, B., & Eger, E. (2016). *Women in tech: The facts*. National Center for Women & Information Technology.
Balsamo, A. (1996). *Technologies of the gendered body: Reading cyborg women*. Duke University Press.
Butler, J. (1990). *Gender trouble: Feminism and the subversion of identity*. Routledge.
Castells, M. (2010). *The rise of the network society*. Wiley-Blackwell.
Cockburn, C., & Ormrod, S. (1993). *Gender and technology in the making*. Sage.
Connell, R. W. (2002). *Gender*. Polity Press.
Crawford, K., & Paglen, T. (2019). *Excavating AI: The politics of images in machine learning training sets*. AI Now Institute.
Derrida, J. (2016). *Of grammatology* (G. C. Spivak, Trans.; Fortieth Anniversary edition). Johns Hopkins University Press.
D'Ignazio, C., & Klein, L. F. (2020). *Data feminism*. MIT Press.
Eubanks, V. (2018). *Automating inequality: How high-tech tools profile, police, and punish the poor*. St. Martin's Press.
Fausto-Sterling, A. (2000). *Sexing the body: Gender politics and the construction of sexuality*. Basic Books.
Gray, K. L. (2018). *Race, gender, and deviance in Xbox live*. Routledge.
Haraway, D. (1991). A Cyborg Manifesto: Science, technology, and socialist-feminism in the late twentieth century. In *Simians, cyborgs and women: The reinvention of nature*. Routledge.
Hochschild, A. R. (1989). *The second shift: Working parents and the revolution at home*. Viking Penguin.
Hill Collins, P. (2000). *Black feminist thought: Knowledge, consciousness, and the politics of empowerment*. Routledge.
Marwick, A. E. (2013). *Status update: Celebrity, publicity, and branding in the social media age*. Yale University Press.
Noble, S. U. (2018). *Algorithms of oppression: How search engines reinforce racism*. NYU Press.
Schwab, K. (2016). *The fourth industrial revolution*. World Economic Forum.
Turkle, S. (1995). *Life on the screen: Identity in the age of the internet*. Simon & Schuster.
Wajcman, J. (2004). *TechnoFeminism*. Polity.

The original version of the book has been revised. A correction to this book can be found at https://doi.org/10.1007/978-3-031-45237-6

Contents

Chapter 1
Technology and Gender Violence: Victimization Model, Consequences and Measures

Sita Rani, Jaskiran Kaur, and Pankaj Bhambri

1.1 Introduction

Rapid technological development is playing a very significant role in the various spheres of society [1]. There is hardly any domain which is not benefitted from this evolution. The use of the internet and various advanced technologies have revolutionized almost every domain of day-to-day life. This technological amalgamation has facilitated life to a great extent. Advanced technologies have a variety of applications in the areas of education, healthcare, agriculture, industry, transportation, smart cities, etc. [2, 3]. Most of the facilitating applications in all these domains are connected in cyberspace and are sharing a huge volume of data. Consequently, there is a huge possibility of cyber threats and unauthorized access to the data. Otherwise, along with advantages, integration of technology has drawbacks also [4, 5]. We can see the negative effect of technology almost everywhere. It affects our day-to-day habits where we are limited to the device-filled life only [6]. It affects both the physical and mental health of an individual. People, especially, children and youth are more addicted to social media where they are exposed to drugs, sexual content, violent videos, etc. Consequently, it results in a lack of empathy among individuals,

S. Rani (✉)
Department of Computer Science & Engineering, Guru Nanak Dev Engineering College, Ludhiana, Punjab, India

J. Kaur
Department of Information Technology, Guru Nanak Dev Engineering College, Ludhiana, Punjab, India

P. Bhambri
Deapartment of Information Technology, Guru Nanak Dev Engineering College, Ludhiana, Punjab, India

D. Mishra et al. (eds.), *Communication Technology and Gender Violence*, Signals and Communication Technology, https://doi.org/10.1007/978-3-031-45237-6_1

and sometimes they are more habitual to cyber/online crimes [7–10]. Some of the common types of cybercrime are:

- Phishing
- Cyber Extortion
- Data Breach
- Identity Theft
- Harassment

Technology-based gender violence is one of the most commonly attempted cybercrimes. It is a complicated episode with destructive results worldwide. With the regular evolution in technology, technology-supported gender-dependent violence is also taking a more advanced and severe form [11]. This technology-equipped violence may cause the spread of personal content (information and/or images) without consent which may lead to serious consequences [12]. In gender-dependent violence, the internet acts as the main weapon of the perpetrators. Different types of technology-equipped gender-dependent violence include threats of damage, spoiling the public image, stalking, etc., many of which are carried out [13–15]. The popular form of gender-based violence includes harassment by an individual or a network of people, sexual abuse using images, distribution of intimate pictures without permission, creepshots, sexploitation, sextortion, sexual assault on public platforms or social media, synthetic media, doxing, defamation, stalking, misrepresentation, impersonation, hate speech, etc. These types of violence cause a physical, emotional, and psychological impact on the victim-survivors [16, 17]. It also impacts the privacy, and safety of the women and girls and may also lead to economic loss. Sometimes, survivors leave the online platforms, cease their public visibility, and restrain their voices.

It has been analyzed that in an intimate relationship, abusive partner tracks their intimate partners, their activities, and communication with the help of technology. Even many women politicians, human right activist, and female journalists usually face different types of threats due to vocalizing about equality or playing a leadership role. People having intersecting marginalized identities are at higher risk of cybercrime, such as people of color, gays, transgender, lesbians, and people with particular types of disabilities. These types of threats/risks cause special safety consideration which involves a threat to privacy and individuality. Technology-facilitated violence causes an unsafe environment for women and LGBTQ+ to move freely in society. It suppresses the voice of women and LGBTQ+ people on social platforms in the era of digital communication.

Gender-based violence abolishes the international human rights of women. Globally, women and girls are targeted to a high degree of violence by their intimate partners than other men [18]. From the literature, women are more frequently targeted for stalking, sexualized violence, and murder irrespective of where they live in this world. Even during public events and gatherings, they face sexual harassment. In war areas too, women and girls face gender-based violence and rape [19]. These are very few examples of everyday occurring violence against women and girls. More precisely, women and girls are the main targets of gender-based violence [20]. But, the study of literature shows that transgender, gay, and men who fall out of the norms of muscularity are also victims of gender-based violence. From statistics, it has been observed that they face harassment, sexual assault, and sometimes physical attacks also reasoned from their expression and gender recognition. It is clear from Fig. 1.1 that there was 23% increase in gender-based violence from

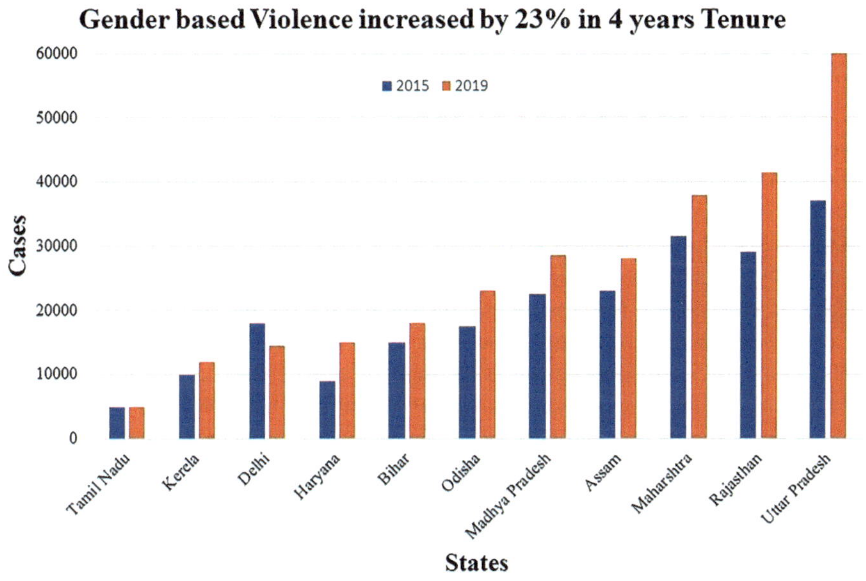

Fig. 1.1 Increase in gender violence since the year 2015 to 2019 in different states of India

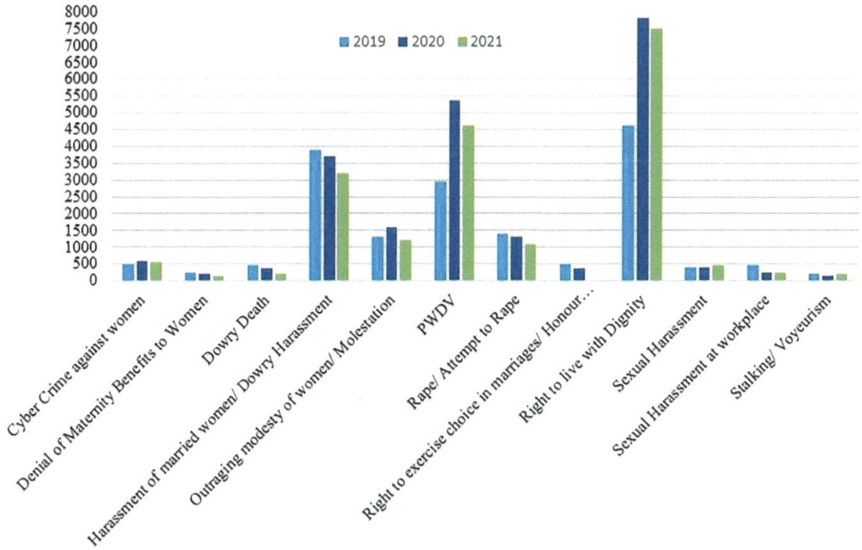

Fig. 1.2 Different types of abuse attempted against women in the years 2019, 2020, and 2021 in India

year 2015 to 2019 in the different states of India. Figure 1.2 presents the variety of abuses attempted against women for the years 2019–2021 in India. As gender-based violence is the biggest threat to gender norms and also causes inequality; consequently, it becomes very important to analyze its mal-effect on the group of victims [21].

1.2 Related Work

In recent years, many authors focused on and carried out their work highlighting the advantages and disadvantages of technology in real-life application domains. A lot of work is also carried out to analyze the role played by modern technologies to encourage gender-based violence [22–24].

In [14], the authors highlighted the usage of technology as media to negotiate sexual identity among today's youth. This agenda is also becoming a central point of discussion during conferences and symposiums as youngsters are using mobile phones and social media platforms for sexting and exchanging personal sexual images and videos to a huge extent. In some countries, a legislative framework to regularize these acts is over-standardized. Even then, the law cannot accurately analyze and regularize the ill effects faced by the victims due to unpermitted development and/or dissemination of sexual content. In this work, the authors focused on the non-consensual development and/or distribution of sexual content as a medium of stalking, harassment, cyber threat, and violence. The main focus of the authors was to investigate the effects of technology-meditated gender-based violence on adult women. It was also concluded that technology is not generating new types of crime but providing new media to accomplish criminal acts. At the same time, these technologies also provide a uniquely new platform to give digital identity to the citizens. The biggest challenge faced is to maintain a balance between freedom of sexual expression and protecting individuals from technology-based gender and sexual violence.

In [13], authors impressed on the agenda of rapidly increasing criminal acts in cyberspace. This work aimed to highlight the role of technology to aid the sexual harassment and violence against women, transgenders, gays, and men for lack of muscular traits, but very little attention is being given in the past to this important sphere. In this article, the authors presented the various types of gender-based threats, sexual assaults, and inequalities conceptualized in cyberspace and caused ill effects. Authors classified technology-based sexual violence into different categories, i.e., unauthorized development and dissemination of visual sexual content, sexual assault images, usage of a communication aid to conduct sexual assault, online stalking and sexual harassment, virtual rape, and gender-focused violent speech. Another important issue, due importance needs to be given to the agenda of gender-based technology-facilitated violent acts such as flaming, cyberbullying, harassment, and trolling against males too. But research advises that girls and women are always major victims of harassment, stalking, assault, and hate speeches in cyberspace. Working to Halt Online Abuse (WHOA) -- an NGO – published (in the year 2011) that the ratio of women victims is very high (approx. 75%) in the reported cases of cyber assault and gender-based violence. These violent acts against women comprised structural, emotional, and physical violence.

In [12], the authors presented sexual violence as a global issue. According to a report presented by the World Health Organization (WHO) in the year 2013, globally 35% of women are experiencing sexual violence either by their partner, a friend, boss, or sometimes even by a family member. From different national-level surveys, it has been concluded that an unbalanced sex ratio and gender in-equalities are a few leading reasons for sexual violence against women. It also evolved from the study that women in the age group of 16–24 are more prone to experience sexual assault. It is also analyzed that the majority of the time this crime is attempted at the hand of a personally or professionally known man. Due to rapid development in the number of smart devices, mobile phones, computers, laptops, tablets, and social media networks, there is a sharp increase in sexual assault, threats, and harms against women. It has also been analyzed that technology is playing a very significant role to aid domestic violence, cyberstalking, dating abuse, and sexual exploitation of teenagers, and children across the globe. "Sexual Violence" is a broad term that is used to describe all types of abusive actions aided by the integration of technology which is called technology-facilitated sexual violence. It addresses a variety of criminal sexually aggressive practices which are aided by the usage of network and communication technologies.

In [25], the authors explored the role of technology in domestic violence against women and girls. The major aim of the study was to analyze the significance of advanced and modern technologies in stalking and other types of cybercrime targeting particular gender. It was stated by the authors that the major technologies to commit this type of crime are social media platforms, mobile phones, and computers. The usual types of crimes executed in the discussed context are harassment, stalking, tracking, etc.

In [26], the authors focused on the contribution of technology in the execution of violent acts by intimate partners. In this work, domestic violence is discussed in a variety of contexts including controlling, cultural and social norms, dynamism in relationships, cultural and structural impositions, etc. The authors concluded their work by emphasizing the use of technology to counter gender-based violence instead of as an aid to perform it.

In [27], the authors discussed the transition in the mode of domestic and sexual violence during the quarantine period of COVID-19. Communal sexual violence changed to technology-based sexual assault. This work was carried out to understand the context of sexual violence in cyberspace. The authors studied the experience of various sexual violence survivors from varying cultural backgrounds. On social media platforms, women are prime targets of sexual violence and assault. This work was concluded with the opinion that being neutral to these types of acts means supporting domestic and gender violence.

In [19], the authors presented a study where mobile technologies and web-based platforms can contribute to improving the mental health of the victims of domestic violence, especially partner abuse. These technologies comprise telehealth services, smartphones, online support groups, etc. It is also highlighted by the authors that psychological disorders, like depression, and PTSD are at the higher end among the victims as compared to the normal population. This review aimed to analyze the effects of partner violence on the mental health of the survivors among all age groups and genders. Authors claimed this study to be the first of its type to analyze the mental health of partner abuse victims using technology.

In [18], the authors expressed their concern about the rapid growth in digital crime. It was highlighted that desired solutions and the legislative frame are not revolutionizing at the same pace. The fundamental aim should be not only to consider the hazards but also to provide counter-solutions. In this work, authors explored both, i.e., technology as an aid to commit a crime and to provide a shield against threats and violence, especially for women and girls. In this work, an in-depth review is carried out in varying aspects of technology-facilitated gender violence, e.g., exploitation, harm, cyberstalking, and harassment of girls and women in common and cyberspace. Another very important aspect, which is being discussed very rarely in literature is the mis-portray of normal behavior as violence in the hour of disagreement to meet their own needs by the other gender.

In [17], the authors stated that globally one-third of the women are experiencing domestic violence either in one form or another where most of them are exploited psychologically, physically, and sexually. In most cases, these acts are accomplished by their partners. In this work, the authors explored the impact of technology-based violence in the domestic sphere. It has been analyzed during the study that the correlation between technology-based gender violence and the domestic front is very less explored in the domain of research. The impact of technology-based gender abuse on the survivors also needs to be analyzed at the next level for suitable guidelines and legal laws. This study aimed to analyze the response of domestic violence counselors to technology-based abuse. The authors used 15 semi-structured interviews and three different themes to gather the response of the experts. Then, this data was used to correlate technology-based assault and other types of domestic violence. It was highlighted in the findings of the study that technology is playing a crucial role to aid domestic violence and can also be used to analyze the degree of the ill effects and the harms.

In [28], the authors revolved their focus around a very interesting question "Is all technology-facilitated abuse violence?". In this work, it is discussed that although most of the physical, psychological, and technology-facilitated violent acts are counted under violence; but, to accurately analyze the complexity of the technological act considering the situation and sequence of factors is important. Some technology-facilitated acts may not meet the threshold of the violence (more appropriately gender-based violence).

In [20], authors explored the types of technology-facilitated gender violence worldwide. The role of various types of technologies including social media, GPS tracking, and AI in the execution of a variety of gender-based violence is discussed in detail. In this work, the authors focused both on quantitative and qualitative research carried out in the literature both on public and private platforms.

Table 1.1 Summarization of previous work: Context, target population, and technologies

Ref. No.	Authors and year	Violence context	Target population	Technology used
[14]	Henry and Powell (2015)	Sexual violence, harassment	Women, girls	Social media platforms, mobiles, email
[13]	Henry and Powell (2015)	Legal policies, harassment	Women, girls	ICT
[12]	Henry and Powell (2016)	Gender-based assault, sexual violence, cyberstalking, harassment.	Women, girls, trans, lesbian	Internet
[25]	Woodlock (2016)	Domestic violence, stalking	Intimate partner	Social media platforms, computers, Mobile phones
[26]	Dragiewicz (2018)	Domestic violence	Intimate partner	Digital platforms
[27]	Jatmiko (2020)	Harassment, domestic violence during COVID-19, online sexual violence during COVID-19	Women	Social media platforms
[19]	Emezue and Bloom (2020)	Depression, psychological disorders, anxiety	Intimate partner	Mobile phones, web-based platforms, wireless platforms
[18]	Marganski and Melander (2021)	Violence in public and private places	Women, girls	ICT, internet
[17]	Fiolet et al. (2021)	Domestic violence	Survivors	Internet, ICT
[28]	Dunn (2021)	International human rights guidelines	Discrimination against women	ICT
[20]	Bailey et al. (2021)	Harassment, assault	Women, girls, gays, transgender	Social Media platforms, artificial intelligence, GPS

Technology-facilitated gender-based violence is explored by researchers in a variety of contexts. The different audience is targeted in different types of criminal attempts using different types of technology in the literature, summarized in Table 1.1.

1.3 Modern Technologies

In the modern era, technology is playing a very significant role in our day-to-day life. Almost every individual is using either one or another kind of modern technology [29, 30]. Several advanced technologies benefit human life in a variety of domains. But few users have started to use this technological development to harm human life. Technology-facilitated gender-based violence is such a bad consequence [31]. Some of the advanced technologies which are used as an aid to attempt violent acts in cyberspace are discussed below:

1.3.1 Information and Communication Technology (ICT)

The architecture, devices, and components that make up a modern computer system are known as ICT [32, 33]. It emphasizes the importance of unified communication protocols and networks, the assimilation of computers and telecommunication services (phone lines and wireless signals), alongside the enterprise software, middleware, repository, and audio-visual components that are required to permit users to access, store, transfer, understand, and maneuver information. ICT is employed in the majority of industries, including banking, e-commerce, agriculture, medicine, education, the military, transportation, and so on.

ICT plays a key role in the execution of technology-facilitated gender-based abuse. Computers, the internet, and communication technology are the most frequently used aids to attempt any violent act in cyberspace [34], as shown in Fig. 1.3. Along with this, mobile phones, tablets, and many social apps are also used as media to attempt gender-based violence.

1.3.2 Social Media Platforms

People use social networking sites and apps to find individuals, chat informally with them, and discover common interests. Social networking platforms enable direct communication between users through groups, connections, and geolocation [35].

Fig. 1.3 Major constituents of ICT

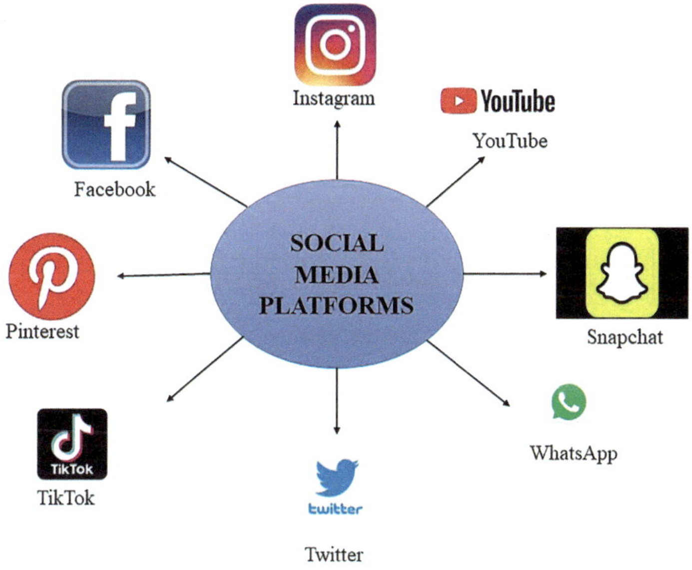

Fig. 1.4 Different social media platforms used to attempt various gender-based violent acts

A few examples of social media platforms include Facebook, Instagram, YouTube, LinkedIn, etc., as shown in Fig. 1.4. These are digital channels for the exchange of information, interests, ideas, and other kinds of expression. Users who connect through these digital services build fully engaging platforms over which people, organizations, and groups can exchange, participate in, co-create, converse about, and change user-generated as well as self-curated material that is published online.

These social media platforms are also used by individuals to perform different types of abusive acts by publishing someone's personal information, defamation, posting negative comments, tracking personal and professional life through posts, etc. [36].

1.3.3 GPS Tracking

Keeping track of the location of an object, also known as geo-tracking, is performed by a navigational tool that is typically mounted on an automobile device, object, individual, or animal [37, 38]. It makes use of the Global Positioning System (GPS) to follow the motion and identify WGS84 UTM's geographic location. GPS tracking systems transmit unique satellite signals, which a receiver processes. The tracking gadget's inbuilt satellite modem, cellular network (GSM/GPRS/CDMA/LTE or SMS), Wi-Fi, or radio can be used to communicate positions to an internet-connected device. The concept of geo-tracking is also used by individuals to track the location of a person or vehicle to perform some violent/abusive act [39]. In complicated relationships, many times women are tracked by their intimate partners to know about their location and keep an eye on their life.

Along with many other technologies like communication networks, disclosure of personal data from various sources also catalyzes online violence.

1.4 Technology-Based Gender Violence Model

It's not only women or girls who are the victims of technology-based gender violence, but the transgender, gay, lesbian, and men with low muscular strength are also targeted either on social media platforms or using other technologies for different types of assaults/crimes, depicted in Fig. 1.5. Harassment, networked harassment, image-based sexual abuse, non-consensual distribution of intimate images, creepshots, sexploitation, sextortion, synthetic media, broadcasting sexual assault, doxing, defamation, public disclosure of private images, hate speech, impersonation threats, stalking, and monitoring are all the major forms of technology-based gender violence [40]. These types of crimes/assaults are committed using different types of advanced technologies, like GPS tracking systems, social media platforms, ICT, mobile phones, tablets, computers, AI, ML, etc. So, the technological development which was aimed to provide better and more convenient life has impacted life badly in this aspect [41]. But it can further be utilized to develop counter-applications and services to address different types of gender-based abuse.

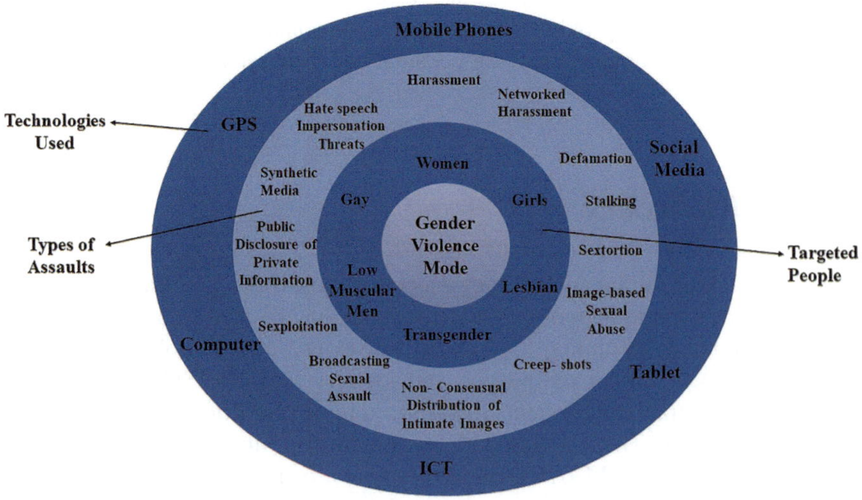

Fig. 1.5 Technology-based gender violence model

1.5 Types of Threats

Technology-based gender violence can be defined as "any act that causes or may cause sexual, psychological, or physical hazards or harms to the girls, women, transgender in private or public using any of the technological development". Even lesbians, gays, and men with low muscularity are also observed as the victim of technology-based violence. Consequently, in our work, we have expanded the interpretation of gender-based violence beyond women and girls to include lesbian, gay, low muscularity men, i.e., any individual experiencing harassment/violence reasoned from the gender/sex and the imposition of benign gender criterion. Although, it is always expected to use technology to make human life better and more convenient; it is being used in several different ways to execute violence in a variety of contexts. The different types of technology-based abuses which are enacted using modern technology are discussed below:

1.5.1 Harassment

Harassment comprises different types of undesired digital communication. It may be very small like a xenophobic comment or an organized and well-planned attack like the Gamergate operation [42]. This act may be attempted by a single person or by a group of abusers. When it is executed by a network of abusers, it is called networked harassment.

1.5.2 Image-Based Sexual Assault

To publicize intimate images (without taking consent) on social media by an ex-partner or any third person is one of the most common types of image-based sexual assault [43]. Along with this, image-based sexual assault comprises several offensive behaviors and comes under the category of "revenge porn".

1.5.3 Creepshots

When any person secretly takes photos or makes videos of another person with any sexual intent is called creepshot. These types of images/videos are used to shame girls and women by posting on online platforms and asking other creepers to comment on them [43].

1.5.4 Sexploitation

It is the act of using someone's sexual material in media for commercial purposes without taking the consent of the concerned person [44].

1.5.5 Sextortion

It is an act in which one person has the sexual images of someone else and on the edge blackmails the person to do something which he/she/they do/do not want [45].

1.5.6 Broadcast/Documentation of Sexual Violence

In this category of abuse, the act of assault is recorded/documented and then disseminated which results in additional harassment to the victim who survived [46].

1.5.7 Publicizing Private Information/Data

Publication of sexual material is a big harassment of women and girls, which is very much clear. But sometimes in many communities with conservative value systems, even the sharing of pic of a woman interacting with a non-family male member, a picture of a woman/girl at a particular place, or wearing a particular type of dress can also put them in trouble and can become the victim of abuse [11].

1.5.8 Doxing

Publication of private data/information, e.g., legal name, passport, driving license, contact information, etc. is known as doxing. Many women, like reporters, and social activists who vocalize against gender inequality are commonly doxed by sexist groups [47].

1.5.9 Defamation

In most nations, defaming simply means publishing wrong information to spoil the reputation of an individual. But in today's era, when everything is available online, it is very easy to check for correct information. It has been observed that nowadays even publishing true private information online can also be dangerous. In many online gender-based attacks, false statements are made public about the sexuality of girls and women to defame them [48].

1.5.10 Stalking

An individual can be stalked in a variety of ways using technology, e.g., regular monitoring of a person for his social media activities, tracking a person using GPS, and the appliance of stalking software on his/her device. It comprises recursive observation and/or dangerous acts which may reason fear. Usually, offensive intimate partners stalk their partners. Many advanced devices/equipment are used to stalk women [49].

1.5.11 Impersonation

It is a type of abuse that can spoil the reputation of an individual and even can cause a plunge. Many times, abusers create fake IDs of leading female personalities to damage their reputations. Even false websites are created to destroy their relationships, work reputation, etc. [50].

1.5.12 Threats

Various types of threats are becoming very common in cyberspace among which online harassment and rape threats are frequently attempted against women. The women who are harassed on social media platforms are highly concerned about their physical safety.

1.5.13 Hate Speech

This category of abuse can initiate/encourage violent acts towards an individual or a group. Any person on the bases of gender, any kind of disability, or religion can be targeted for hate speech. The criminals can target any woman/girl due to their position, identity, looks, etc. This may discourage women/girls to be very active on social media platforms [51].

1.6 Who Gets Troubled?

From statistics, women and girls are the major victims of technology-based gender violence. Along with this, many other groups of people are also targeted for this kind of abuse/violence. Transgender, people with different sexual orientations or any other kind of disability/class can also become the victims of technology-based violence. In this section, our main aim is to analyze the different classes of the population and the context in which they are targeted for technology-based violence.

1.6.1 Intersectional Equivalence Aspects

Technology-based gender violence is caused by racism, discrimination, transphobia, etc. It is very common for women and girls to be targeted for this abuse by anti-feminists and sexists in cyberspace cantered around their sexual orientation, community, race, any type of disability, expression of gender, etc. It has been also analyzed that an individual's intersectional identity aspects will affect their online experience in terms of types of attacks and degree of violence planned/executed on them. For example, even in the case of technology-based attacks, black women are targeted more in comparison to white women.

1.6.2 Women Performing Lead Roles

It has been observed that women are kept on the back-front to allocate lead roles worldwide. In the modern era, there is an urgent requirement for gender diversity in many of the professions/domains such as journalism, politics, and lead roles in many other professional domains. But this aspect is suppressed when actively participating women become victims of technology-based gender violence. Technology-based gender violence stops them from working effectively and also becomes the reason to keep them away from lead roles, especially when they are vocal against any type of discrimination or social inequality. It restricts women and girls to engage-in in all domains of day-to-day life.

1.6.3 Violence by Intimate Partner

Women are also targeted for different types of abuse by their intimate partners. They become the victim of technology-facilitated violence. From the study of the literature, it has been observed that almost 2/3rd of the abuses are planned and executed by their current or ex-intimate partner. Usually, their life is tracked through their mobile devices, social media accounts, etc. It makes any woman in fear and threat as they feel that they are always under observation. Abuse by an intimate partner may also be in the form of physical violence or domestic violence.

1.7 Types of Harms

Technology-facilitated gender violence may be planned against an individual or it may have systemic impacts. When we talk about an individual victim, he/she may experience privacy invasion, threat to autonomy, psychological threat, reputational attack, fear, professional risks, and economic harm. It causes inequality and discriminant behavior toward women/girls and transgender which limits their freedom and restricts their human rights [20]. In this way, technology-based gender violence can cause a variety of harm to the victims, as shown in Fig. 1.6.

1.8 Law to Administer Technology-Based Gender Violence

Technology-facilitated gender-based violence is considered a comparatively new domain in India. However, several organizations have started to work to address the various issues related to this issue. Several organizations, like UN Women, Women's Fund Asia, Omidyar Network, Mozilla Foundation, Tata Trust, Amplify Change, and many more are financing the various initiatives

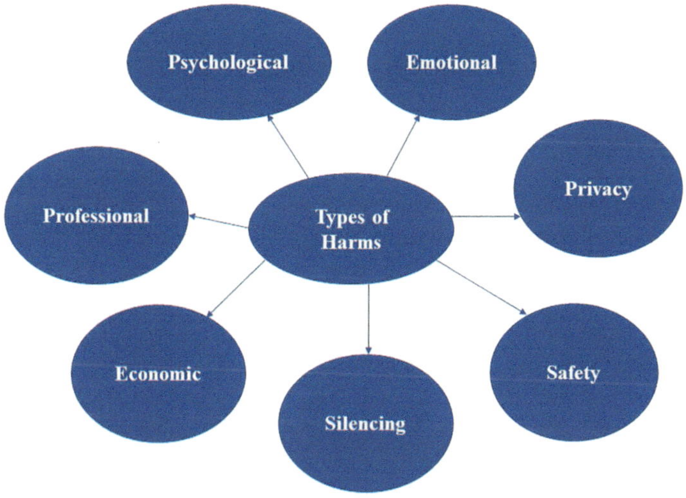

Fig. 1.6 Types of harms: Victims of gender violence

taken to protect digital rights, control technology-facilitated gender-based violence, and provide digital safety to the people [52, 53].

Many NGOs have also taken initiatives to make the users digitally literate to not be affected by online violence. They have also started to create better frameworks to report and monitor the various abusive activities attempted in cyberspace.

The Government of India is also taking many initiatives to aware people of technology-facilitated gender-based violence and to administer the issue. Many cyber cells have been activated by looking at the seriousness of the domain. Reporting mechanisms are also very efficient. A few states are using Nirbhaya Fund under Safer City initiatives.

1.9 Precautionary Measures

Technology-facilitated gender-based violence is one of the major problems faced by the society. Although, in this technological era, it is not possible to stop it completely, there are a few precautionary measures that can help to limit it to a huge extent, discussed below [54]:

- To enhance digital literacy among the users.
- To make strict laws and their enforcement.
- To provide better support and resources for the survivors.
- To administer social issues and cultural norms which cause violence.
- To work in collaboration with various social media sites and companies for content moderation.
- To improve the system to report and track the issues.

1.10 Conclusions and Future Scope

From the literature, it has been observed that technology is acting as a catalyst to execute violent acts and harassment in society. It has become an international issue that requires extra attention. Women/girls, transgender, and even muscularly weak men are common victims of various types of abuse. Women, who are in leadership roles and more vocal about discrimination and their rights, are the main targets in technology-facilitated violence on social media platforms. They also become the victim of domestic violence and abuse by intimate partners. There is a class of gender-based abuse, like harassment, publicizing private information and intimate pictures, stalking, threats, hate speech, threats, etc. The major harms caused are safety issues, invasion of privacy, psychological risks, economic loss, reputation spoiling, etc.

Although in this technological era, it is not possible to control technology-facilitated gender-based violence completely but many NGOs, private bodies, and government organizations have started to sensitize users to digital literacy. The government has started to make strict laws and their enforcement in the domain. Many systems to report violent acts and track have been developed and implemented.

References

1. Bibri, S. E., & Krogstie, J. (2017). On the social shaping dimensions of smart sustainable cities: A study in science, technology, and society. *Sustainable Cities and Society, 29*, 219–246.
2. Kumar, A., Gupta, P. K., & Srivastava, A. (2020). A review of modern technologies for tackling COVID-19 pandemic. *Diabetes & Metabolic Syndrome: Clinical Research & Reviews, 14*, 569–573.
3. Tanwar, R., Chhabra, Y., Rattan, P., & Rani, S. (2022). Blockchain in IoT networks for precision agriculture. In *International Conference on Innovative Computing and Communications: Proceedings of ICICC 2022, Volume 2* (pp. 137–147). Springer.
4. Kumar, P., et al. (2022). Verifiable, secure mobile agent migration in healthcare systems using a polynomial-based threshold secret sharing scheme with a blowfish algorithm. *Sensors, 22*, 8620.
5. Dhanalakshmi, R., Anand, J., Sivaraman, A. K., & Rani, S. (2022). IoT-based water quality monitoring system using cloud for agriculture use. In *Cloud and Fog Computing Platforms for Internet of Things* (pp. 183–196). Chapman and Hall/CRC.
6. Alghamdi, Y. (2016). *Negative effects of technology on children of today* (Vol. 10). Oakl Univ.
7. Datta, P., Panda, S. N., Tanwar, S., & Kaushal, R. K. *A technical review report on cyber crimes in India* (pp. 269–275). IEEE.
8. Rani, S., Kataria, A., & Chauhan, M. (2022). Cyber security techniques, architectures, and design. In *Holistic approach to quantum cryptography in cyber security* (pp. 41–66). CRC Press.
9. Rani, S., Kataria, A., Kumar, S., & Tiwari, P. (2023). Federated learning for secure IoMT-applications in smart healthcare systems: A comprehensive review. *Knowledge-Based Systems, 274*, 110658.
10. Kataria, A., Agrawal, D., Rani, S., Karar, V., & Chauhan, M. (2022). Prediction of blood screening parameters for preliminary analysis using neural networks. In *Predictive modeling in biomedical data mining and analysis* (pp. 157–169). Elsevier.
11. Harris, B. A. (2020). Technology and violence against women. In *The emerald handbook of feminism, criminology and social change*. Emerald Publishing Limited.
12. Henry, N., & Powell, A. (2018). Technology-facilitated sexual violence: A literature review of empirical research. *Trauma, Violence, & Abuse, 19*, 195–208.
13. Henry, N., & Powell, A. (2015). Embodied harms: Gender, shame, and technology-facilitated sexual violence. *Violence Against Women, 21*, 758–779.
14. Henry, N., & Powell, A. (2015). Beyond the 'sext': Technology-facilitated sexual violence and harassment against adult women. *Australian & New Zealand Journal of Criminology, 48*, 104–118.
15. Shelby, R. (2021). Technology, sexual violence, and power-evasive politics: Mapping the anti-violence sociotechnical imaginary. *Science, Technology, & Human Values*, 01622439211046047.
16. Dunn, S. (2020). Technology-facilitated gender-based violence: An overview. In *Centre for International Governance Innovation: Supporting a safer internet paper*, 2020.
17. Fiolet, R., Brown, C., Wellington, M., Bentley, K., & Hegarty, K. (2021). Exploring the impact of technology-facilitated abuse and its relationship with domestic violence: A qualitative study on experts' perceptions. *Global Qualitative Nursing Research, 8*, 23333936211028176.

18. Marganski, A. J., & Melander, L. A. (2021). Technology-facilitated violence against women and girls in public and private spheres: Moving from enemy to ally. In *The Emerald International handbook of technology facilitated violence and abuse*. Emerald Publishing Limited.
19. Emezue, C., & Bloom, T. L. (2021). PROTOCOL: Technology-based and digital interventions for intimate partner violence: A meta-analysis and systematic review. *Campbell Systematic Reviews, 17*, e1132.
20. Bailey, J., Henry, N., & Flynn, A. (2021). Technology-facilitated violence and abuse: International perspectives and experiences. In *The Emerald International handbook of technology facilitated violence and abuse* (pp. 1–17). Emerald Publishing Limited.
21. Aubert, A., & Flecha, R. (2021). Health and Well-being consequences for gender violence survivors from isolating gender violence. *International Journal of Environmental Research and Public Health, 18*, 8626.
22. Bali, V., Bali, S., Gaur, D., Rani, S., & Kumar, R. (2023). Commercial-off-the shelf vendor selection: A multi-criteria decision-making approach using intuitionistic fuzzy sets and TOPSIS. *Operational Research in Engineering Sciences: Theory and Applications, 6*, 1–18.
23. Puri, V., Kataria, A., Solanki, V. K., & Rani, S. (2022). AI-based botnet attack classification and detection in IoT devices. In *In 2022 IEEE international conference on machine learning and applied network technologies (ICMLANT)* (pp. 1–5). IEEE.
24. Rani, S., Bhambri, P., Kataria, A., & Khang, A. (2022). Smart city ecosystem: Concept, sustainability, design principles, and technologies. In *AI-centric smart city ecosystems* (pp. 1–20). CRC Press.
25. Woodlock, D. (2017). The abuse of technology in domestic violence and stalking. *Violence Against Women, 23*, 584–602.
26. Dragiewicz, M., et al. (2018). Technology facilitated coercive control: Domestic violence and the competing roles of digital media platforms. *Feminist Media Studies, 18*, 609–625.
27. Jatmiko, M. I., Syukron, M., & Mekarsari, Y. (2020). Covid-19, harassment and social media: A study of gender-based violence facilitated by technology during the pandemic. *The Journal of Society and Media, 4*, 319–347.
28. Dunn, S. (2021). Is it actually violence? Framing technology-facilitated abuse as violence. In *The Emerald international handbook of technology facilitated violence and abuse*. Emerald Publishing Limited.
29. El Morr, C., & Layal, M. (2020). Effectiveness of ICT-based intimate partner violence interventions: A systematic review. *BMC Public Health, 20*, 1–25.
30. Rodríguez-Rodríguez, I., Rodríguez, J.-V., Elizondo-Moreno, A., Heras-González, P., & Gentili, M. (2019). Towards a holistic ICT platform for protecting intimate partner violence survivors based on the IoT paradigm. *Symmetry, 12*, 37.
31. V. M. De Stefano, I. Durri, C. Stylogiannis, and M. Wouters, "System needs update": Upgrading protection against cyberbullying and ICT-enabled violence and harassment in the world of work, 2020.
32. Zaidi, A. U., Fernando, S., & Ammar, N. (2015). An exploratory study of the impact of information communication technology (ICT) or computer mediated communication (CMC) on the level of violence and access to service among intimate partner violence (IPV) survivors in Canada. *Technology in Society, 41*, 91–97.
33. Backe, E. L., Lilleston, P., & McCleary-Sills, J. (2018). Networked individuals, gendered violence: A literature review of cyberviolence. *Violence and Gender, 5*, 135–146.
34. Prendes-Espinosa, M.-P., García-Tudela, P.-A., & Solano-Fernández, I.-M. (2020). Gender equality and ICT in the context of formal education: A systematic review. *Comunicar, 28*, 9–20.
35. Tripathi, V. (2017). Youth violence and social media. *Journal of Social Sciences, 52*, 1–7.
36. Kavanagh, E., Litchfield, C., & Osborne, J. (2019). Sporting women and social media: Sexualization, misogyny, and gender-based violence in online spaces. *International Journal of Sport Communication, 12*, 552–572.
37. Erez, E., Ibarra, P., Bales, W., & Gur, O. (2012). *GPS monitoring technologies and domestic violence: An evaluation study*. U.S. Department of Justice, National Institute of Justice.

38. Kennedy, K. M., & White, C. (2015). What can GPs do for adult patients disclosing recent sexual violence? *British Journal of General Practice, 65*, 42–44.
39. Rout, A., Nitoslawski, S., Ladle, A., & Galpern, P. (2021). Using smartphone-GPS data to understand pedestrian-scale behavior in urban settings: A review of themes and approaches. *Computers, Environment and Urban Systems, 90*, 101705.
40. Fulu, E., & Miedema, S. (2015). Violence against women: Globalizing the integrated ecological model. *Violence Against Women, 21*, 1431–1455.
41. O'Neil, J. M., & Harway, M. (1997). A multivariate model explaining men's violence toward women: Predisposing and triggering hypotheses. *Violence Against Women, 3*, 182–203.
42. Finn, J. (2004). A survey of online harassment at a university campus. *Journal of Interpersonal Violence, 19*, 468–483.
43. Call, C. (2021). Perceptions of image-based sexual abuse among the American public. *Criminology, Criminal Justice, Law & Society, 22*, 30145.
44. Coopey, L. (2022). Sexual violence and smallfolk: The exploitation of the sex worker in. In *The forgotten victims of sexual violence in film* (pp. 47–62). Television and New Media: Springer.
45. Eldén, Å., Calvo, D., Bjarnegård, E., Lundgren, S., & Jonsson, S. (2020). *Sextortion: Corruption and gender-based violence*. Expert Group for Aid Studies (EBA).
46. Scully, D. (2013). *Understanding sexual violence: A study of convicted rapists*. Routledge.
47. Douglas, D. M. (2016). Doxing: A conceptual analysis. *Ethics and Information Technology, 18*, 199–210.
48. Thomas, K., et al. (2021). SoK: Hate, harassment, and the changing landscape of online abuse. In *2021 IEEE symposium on Security and Privacy (SP)* (pp. 247–267).
49. Al-Rahmi, W. M., Yahaya, N., Alamri, M. M., Aljarboa, N. A., Kamin, Y. B., & Saud, M. S. B. (2019). How cyber stalking and cyber bullying affect students' open learning. *IEEE Access, 7*, 20199–20210.
50. Leyton Zamora, C., Boddy, J., O'Leary, P., & Liang, J. (2021). Culturally and linguistically diverse (CALD) Women's experiences of technology-facilitated violence: An intersectional approach. In A. Powell, A. Flynn, & L. Sugiura (Eds.), *The Palgrave handbook of gendered violence and technology* (pp. 115–134). Springer.
51. Saresma, T., Karkulehto, S., & Varis, P. (2021). Gendered violence online: Hate speech as an intersection of misogyny and racism. In M. Husso, S. Karkulehto, T. Saresma, A. Laitila, J. Eilola, & H. Siltala (Eds.), *Violence, gender and affect: Interpersonal, institutional and ideological practices* (pp. 221–243). Springer.
52. Rushing, S. C., & Stephens, D. (2011). Use of media technologies by native American teens and young adults in the Pacific northwest: Exploring their utility for designing culturally appropriate technology-based health interventions. *The Journal of Primary Prevention, 32*, 135.
53. Snaychuk, L. A., & O'Neill, M. L. (2020). Technology-facilitated sexual violence: Prevalence, risk, and resiliency in undergraduate students. *Journal of Aggression, Maltreatment & Trauma, 29*, 984–999.
54. Lumsden, K., & Morgan, H. (2017). Media framing of trolling and online abuse: Silencing strategies, symbolic violence, and victim blaming. *Feminist Media Studies, 17*, 926–940.

Chapter 2
Implication of Technology to Curb Digital Violence Among Women on Social Media Platforms

Deepanjali Mishra and Mangal Sain

2.1 Introduction

The recent case which happened a few days back has taken people by storm. In this case, a young medical female student in Karnataka faced harassment from another male medical intern called Saif after which she injected herself with a harmful substance [1]. The harassment she faced was unbearable. Violence against women occurs in various forms which is known as gender-based violence. They could be physical as well as mental and could lead to severe injuries, suicides, and inflicting harm on self and others. It takes place not only in the four walls of the homes, but also it is rampantly taking place in the workplace. What's most horrifying is that it is not country-specific, but it has become a global phenomenon. Cyberfeminism is one that makes the relationship between cyberspace, Information and Communication Technology (ICT), and the internet in general more prominent. Living in the twenty-first century we still haven't been able to do justice in giving parity to women as a gender in the cyberspace. Hence empowering women for their emancipation in this platform has become the need of the hour. There is a concept of masculinization in which technology is considered as masculine and that is the general belief and hence women are compared to tech-savvy men [2]. According to research, the internet is falling out of women users because of cyber harassment, cyber bullying, and cyber teasing. With services like online payments (including Universal Payment Interface (UPI) in every other transaction (in India)), digital banking, digital media, IOT (Internet of Things) on the rise, there is IPV (Intimate Partner Violence), revenge porn (non-consensual sharing of intimate photos), rape videos, morphed

D. Mishra (✉) · M. Sain
School of Humanities, KIIT University, Bhubaneswar, India

Computer Science Division, Dongseo University, Busan, South Korea

D. Mishra et al. (eds.), *Communication Technology and Gender Violence*, Signals and Communication Technology, https://doi.org/10.1007/978-3-031-45237-6_2

images, child pornography, etc., acting conspicuously to its detriment. Women are having to face the brunt of such blatant misuse of the cyberspace. The COVID-19 pandemic has rendered a lot many people jobless and with this there has been a marked increase in criminal activities on the internet that needs serious acknowledgement. People have resorted to scam calls and messages for siphoning money and through other methods of social engineering to befool the most naive to invest money in the name of Ponzi schemes [3]. It has been found that women, especially housewives are being trapped in numbers because of lack of technical know-how in certain cases and lack of exposure in most.

Cyber bullying is on the rise and women are the majority stakeholders to be on the rough end of the stick and hence cyber bullying and cyber teasing have to be taken seriously because they are one of the major reasons for spoiling the mental state and mindset of women who surf the internet for some kind of solace in search of equality if not opportunity which unfortunately is still a farfetched idea. Cyber bullying however is very less talked about and its existence is a lesser-known fact. People have a tendency of trolling and shaming others on social media in the quest to look smart while demeaning and undermining the mental, physical, and emotional state of the recipient. There cannot be a check on conversations owing to the freedom of free speech and expression but definitely on certain verbatims through the intelligent and resourceful use of machine learning and artificial intelligence.

There is an intrinsic relation of cyber security with gender violence. Women have been at the receiving end of the stick facing the brunt of gender discrimination even in the virtual world where their extortion and harassment is seen often in opposition to the pledge of the internet for neutrality and equality. They have been catcalled, body shamed, and constantly reminded of their position in the society which promised them a level playing field, if not anything else, at the least. Hence, a solution--oriented approach is the need of the hour as has been elucidated in the paper, connecting the dots and linking every other issue in cyber security to master the art of living in the digital space, deriving optimum output from the same without being masqueraded by the pretentious elements with ulterior motives.

2.2 Review of Literature

Many researches have been conducted on the impact of gender violence and social media. Social contacts in the physical world or in social media can have a strong influence on the attitude of individuals. An extensive body of literature has studied how social media's exposure can influence an individual's psychological states. Other work has explored the influence of content creation on social media attitude, such as retweeting, replying, or favoriting, for example, in the context of the Twitter platform. For instance, Leavitt et al. 2009 classified user's influence into two types: content-based and conversation-based. This work concluded that influential people such as celebrities were better at starting conversations on social media while news outlets content resulted in more retweets. The ultimate form of influence is to promote collective action via social networks; this was visible in the Black Lives Matter (BLM) movement and the Arab Spring. On the theoretical end of studying influence and factors that promote users to endorse certain campaigns, points of view, or products, lie the theories of Influence Maximization and Contagion. Influence Maximization is the problem of finding a set of nodes in a network that maximizes the spread of an idea or campaign.

2.3 Technology and Digital Violence

Technology has played a major role in aggravating the gender-based violence. There are various crimes that happen to women on social media. Technology-facilitated GBV occurs worldwide. A recent study that was conducted in Asia says that online harassment does take place among women and it can shoot up to 40 percent among adults and up to 33 percent among adolescents. In the USA about 45 percent of women confessed that they have faced online harassment which can range from the culprit's behavior, stalking, eve-teasing, cyber bullying, hate messages defamation, and sending abusive messages. Sometimes online form of abuses can lead to offline violence where the victim may cause harm to self or the culprit may harm her. One of the important features of technology-based crimes is that the culprit can be anonymous while committing the crimes which can pose to be more fatal for the victim. Privacy invasion leading to infiltration into someone's personal space has become a thing very common now. This causes hate messages in the form of comments, or trolling caused by multi users.

There are various forms of digital gender violence. Some of them are as follows:

2.3.1 Revenge Violence

This is a grave version of sextortion, a portmanteau or frankenword for "sexual" and "extortion" which is the non-consensual sharing of photos or videos while sextortion basically being the threat of sharing photos whether they are shared or not. With the advancement of social media and with farrago of platforms having come up for exchange of conversations through texts, emoticons, photos, and videos, extortion through photos and videos with explicit content being shared in these platforms has become very common. Various dating applications and social media platforms will always try to be better in a competitive world where each one of them is striving to be the best but that should not make users take things lightly while operating and using these applications for sharing personal documents, photos, and videos. The most vital element of sextortion is the use of blackmail and the power exerted by the offender on their partner, coercing their compliance into a certain action which in itself is a greater crime [4]. Such activities have become the new normal and have resulted in the suppression of voices that have not been able to free themselves from the shackles of toxicity and such people normally make the headlines of news dailies for having given up their lives, trying to save the so-called image and repute of their families [5]. Revenge porn does not only happen locally or with people who have prior acquaintances but also occurs on a large scale transnationally as well. Such activities have been promulgated by dating applications and platforms where the initial acquaintance is mostly through sharing of photos, even with the prior knowledge of the other person being miles apart or time zones apart. Such has been the fad or the trend that the conscious of the layman has taken a backseat. Therefore, such people need to understand that the sharing of such intimate details may lead to morphed images being created and being used in rape videos as well. Legislations for the same are also very tricky as establishing precedent for "no consent" becomes very difficult as these are intricate matters that have a very delicate line of action through which matters have occurred in real time and to showcase the same in the form of proof in the court of law, however right one may become an arduous task. We also need to make ourselves aware of these nefarious activities on the rise and take steps to educate the younger ones to not fall into such traps and be able to make optimum use of the resources available to them.

2.3.2 *Cyber Bullying*

Cyber bullying, although very less talked about, is a lesser-known fact anyway. Trolling and shaming people in the name of fame and attention projects a very parochial mindset and because of no stern legislation for cyber bullying so to speak there is undue advantage taken on the internet in the name of freedom of speech and expression. It can have psychological effects on the recipient mostly because of the anonymous nature of messaging and because it's done in an open platform where people from all spheres and aspects of life are associated [6]. The quest to maintain a social image, although one may be diametrically opposite to it in reality helps them be a pretentious form of themselves that acts to their detriment when they are not able to fulfil their expectations even on their social media platform. This creates a sense of low morale and low self-esteem often leading to depression and anxiety. Bullies in real life can be avoided but then cyber bullies are everywhere because every other person is on the internet and therefore is connected to the world of social media in one way or the other. The act of undermining or demeaning someone becomes a convention so much so that it tends to have a pernicious influence on the youth of today often resulting in heated exchange of words and culminating into a war of words till the point in time where there are too many against a single entity trying to prove their point and make a statement and thus singling out, isolating and secluding that one person who ends up having an existential crisis. All this results in the imbalance of the mental, physical, and emotional state of the person at the receiving end of the stick. Therefore, putting a check on the same may not be possible as filtering out every conversation would not be practical but then the scrutiny of certain verbatims through the intelligent use of machine learning and artificial intelligence can definitely go a long way in solving the current problem at hand.

2.3.3 Violence on Women Through Social Media

There is an intrinsic relation between cybersecurity and gender violence. Women have been facing the brunt of gender discrimination including their extortion and harassment in the virtual world where there were expectations of a level playing field but turned out be no different than the real world. They have been catcalled and slut shamed exacerbating their already poor condition in the society. There is multifarious violence against women in the cyber world such as those of sexual harassment, revenge porn, online solicitation, rape threats, gender-based hate speech, unsolicited nude images, doxing, etc. Most of it is seen in cases of misogynists who target those women who are feminists and try and blackmail them from taking an unrelenting stand to further their own cause. Technology should have always acted as a facilitator and been gender neutral at the least but then it is technology that has rather made the access to political knowledge and avenues for civic management limited in the case of women which speaks volumes of the ingrained mindset and mentality and hence only showing that the same issue that women face offline have only been reified and manifested online. There are classical examples of women who have single handedly taken up such causes to fight for their entire sex and de--fang and dodge all the criticisms hurled at them but then there must be support from the technology front as well otherwise the holier than-thou attitude of the society will keep shoving responsibilities on women while the world watches in silence. Among all this that has been happening, the individual responsibility is a priority when it comes to technology-facilitated gendered violence (TFGV) as there are numerous ways of preventing such mishaps or violence for women on the internet if precautionary measures for the same are taken like that of desisting social media use, filtering unsolicited content and abuse through privacy settings and simply blocking. At the same time community responsibility cannot be ignored, one that starts with educating the young mass and policing the elderly as and when required so that the essence of a civilized society is perpetuated.

Gender-based violence, or GBV as it is commonly called, is mostly targeted against women. However, it cannot be denied that there are cases where it's the other way round. Many men have fallen prey to gender-based violence but statistics show that such cases are very less compared to violence inflicted on women. It is mainly due to the fact that women are ignorant or may not be too much conversant to internet compared to men. Undoubtedly it gets propagated through various social media platforms which are available on internet like Twitter, Facebook, and even Linkedin [7]. There are some platforms like Linkedin that have maintained transparency regarding visibility and inclusion, yet more needs to be done. It also has taken an initiative to protect the interests and privacy of transgender society, which is secluded from the mainstream media but such platforms and turning into hubs for the netizens to post comments which are offensive in nature in the public as well as personal chatbox invading the privacy of the users.

This version of GBV, against transsexual people, is mirrored across platforms, including non-personal platforms such as LinkedIn. On a professional platform like LinkedIn, it is harder to find trans persons. Those who are prevalent usually have a political/activism platform outside of their professional career. Though, as a platform, LinkedIn does advocate for trans visibility and inclusion, the community remains a minority even online, where often, their mere presence becomes a place for people to comment on offensive and transphobic statements both on the publicly visible parts of the platform and in private interactions. LinkedIn, a platform for professional connection and communication, has also seen an increase in harassment reports and reports of inappropriate private messages, to counter this, the platform has also begun to use AI and machine learning tools to grade users by their use and interaction and reduce harassment at its point of origin.

Though a step forward, this step increases the responsibility of those on the receiving end to make decisions and report the harassment. This brings up primarily two issues: The lack of control of users outside the platform and the perceived para--social relationships may hinder reporting.

An example of users lacking control over GBV outcomes can include deep fake technology. Often used to create false pornographic content, has implications of social consequence almost immediately, if used against marginalized genders. The law, so far, both in India and globally, is currently ill-equipped to contain the spectrum of deepfake use, especially when it comes to forms of sexual harassment. Though the IPC does currently protect against receiving unwarranted obscene images and videos, the technology allows for many conduits for the offending perpetrators, ranging from freedom of speech to even copyright laws. The solution would be to create a comprehensive regulation to overlook the creation and circulation of non-consensual videography and pornography.

GBV in private spaces is often a result of para-social relationships. The perception of these relationships affects not just the offender but also the victims, with some feeling guilty about reporting a "connection" and some not viewing the harassment as offensive due to a previous or underlying relationship. The process of reporting puts the onus on the marginalized genders to ensure those sending lewd messages are monitored, removed, or banned. This subjective policing prevents many disenfranchised by status or position, especially in terms of professional positions, from reporting instances that they may deem as insignificant forms of harassment.

The process of reporting puts the onus on the marginalized genders to ensure those sending lewd messages are monitored, removed, or banned. The guise of safety and concern has been used to monitor and restrict, and that of networking, friendship, and romance to approach women across platforms.

2.3.4 The Existing Scenario

At present, there are no laws or regulations at the state level or at the international level which can be used by women to seek protection from the violence that is being inflicted on them through social media. However looking into the alarming rise in the number of cases, law makers have started to understand the need of framing norms. These norms prove to be less effective and efficient because the perpetrators of crimes prove to be more shrewd and they can easily escape after committing the crimes. Therefore, those norms which are social media platform-specific need to be formulated, evaluated, and implemented. It is also very important that strong alternatives of these norms can be developed [8]. Various distinguished law makers, industrialists, corporate personnel, technology experts, and police personnel can form a group while preparing these laws.

2.4 Using Technology to Curb Gender Violence

Some of the case studies will be analyzed to show how technology could be successfully used to curb crimes against women.

2.4.1 Understanding Social Media

A very common feature that was found among women in Uganda is that women keep two SIM cards to avoid domestic violence which is quite interesting. One SIM was used at home and the number was known to their husbands whereas the second SIM number was shared only with those people who could generate doubts if they were to call in presence of their husbands to raise alarm at the office and someone could come immediately for help. A very common feature that was found among young girls in Africa is that they spent up to 6 hours a day on their mobile phones to understand its technology irrespective of their background [9]. They would also understand to educate themselves about the methods of chatting, texting and messages so that they would understand the techniques to protect themselves from being victim of technology-aided violence.

2.4.2 Creating Awareness on Social Media Using Digital Story Technique

One of the very innovative techniques used by the teenagers in South Africa to create awareness about gender violence and its prevention is through Digital Story Telling which emphasizes the issues and the roles and rights of men and women among their communities. For example, a girl named Tokozile created a digital profile of her friend who was raped by her boyfriend after taking the consent of the victim [10]. It was through this method, people from all vicinity came to know about the incident as it gave opportunity to talk about such humiliating incidents to make people realize the experience of the victim what she might have gone through. After this incident, Tokozile's friend has become a teacher now who is creating awareness among other teenage girls about rape. In Uganda, there are many women empowerment organizations, who use internet to highlight violence against women like rape, molestation, victimization, and harassment.

2.4.3 Using ICT Tools for Redressals of Gender Violence

Internet and technology has helped people across the world to use it as tools of redressals. One such example is that of Margaret who was repeatedly abused and beaten by her husband. Not only that, he even murdered her sister and mother. No matter how much she tried to file case, it was not possible due to corruption existing in Sri Lanka [11]. However, some NGOs and women rights group came to know about it and they recorded the case using media technology and this recording was uploaded online which finally caught the attention of the authorities and the case was taken up. It was granted urgency hearing and justice was finally given to the victim.

2.4.4 Using ICT Tools for Recognition

A very much (in)famous case that had caught international attention was that of a woman who was flogged by few men in public in Pakistan. Her only fault was that she was seen with a man who was not her immediate cousin [12]. This incident occurred in the tribal area of Swat valley. It was in March 2009, when this video was uploaded on YouTube which caused severe outrage among the viewers. It showed a woman who was covered all over, was on the ground pleading for mercy while few men were brutally flogging her and she was crying with pain [13]. After this various talk shows were organized on this incident which discussed about the talibanization in the Pakistani society.

2.5 Conclusion

Thus it can be concluded that cyber media is one of the major causes of violence across the genders and it is more rampant among the teenaged population. Apart from that Covid-19 pandemic which restricted users to switch over to online mode also could be another reason for the cause of rise in cyber violence [14]. It is only through the use of cybersecurity measures that it could be lessened to some extent. Practical implication of cyber security in the form of e-learning is more important than the theoretical aspects because it is only through practical applications that an user is able to develop new strategies to curb the violence by being able to access the devices and understand its applications. It needs to be understood that cybersecurity improves the security of people of all gender identities and expressions, as well as international peace and security. The ultimate conclusion is that these two levels of security cannot be separated or could be treated as a different entity.

2.6 Scope of Further Research

This topic of how E-learning can be used to sensitize women to curb digital violence is quite exploring [15]. It has a vast scope of further research. Various innovative tools of research, research methodologies, and research models could be developed and devised which could help women to protect themselves from falling into prey of being victimized on social media.

References

1. Gender based violence and unwanted sexual behavior in Canada, 2018. Available at https://www150.statcan.gc.ca/n1/daily-quotidien/191205/dq191205b-eng.htm ii. European Union (2018). Cyber violence and hate speech online against women. Available at https://www.europarl.europa.eu/RegData/etudes/STUD/2018/604979/IPOL_STU(2018)604979_EN.pdf iii. Pew Research Center (2017). Online Harassment 2017. Available at https://www.pewresearch.org/internet/2017/07/11/online-harassment-2017/iv.European Union Agency for Fundamental Rights (2014). Violence against women: an EU-wide survey. Available at https://fra.europa.eu/sites/default/files/fra_uploads/fra-2014-vaw-survey-main-results-apr14_en.pdfv. Digital Rights Foundation (2017). Measuring Pakistan Women's Experiences of Online Violence. Available at https://digitalrightsfoundation.pk/wp-content/uploads/2017/05/Hamara-Internet-Online-Harassment-Report.pdf
2. African Development Bank Group. (2016). *Minding the gaps: Identifying strategies to address gender-based cyberviolence in Kenya.*
3. https://www.afdb.org/fileadmin/uploads/afdb/Documents/Generic-Documents/Policy_Brief_on_Gender_Based_Cyber_Violence_in_ Kenya.pdf. 2 Burton, P.
4. Mutongwizo, T. (2009). Inescapable violence: Cyber bullying and electronic violence against young people in South Africa. *Centre for Justice and Crime Prevention., 8*, 1–12.
5. De Lange, M., & von Solms, R. (2012). An e-safety educational framework in South Africa. In *Southern Africa Telecoms and Network Applications Conference (SATNAC).*
6. http://www.satnac.org.za/proceedings/2012/papers/3.Internet_Services_End_User_Applications/53.pdf. Duggan, M. Online Harassment. Pew Research Center. 2014.
7. http://www.pewinternet.org/2014/10/22/online-harassment/. Ojanen, T. T., Boonmongkon, P., Samakkeekarom, R., Samoh, N., Cholratana, M., & Guadamuz, T. E. (2015). Connections between online harassment and offline violence among youth in Central Thailand. *Child Abuse & Neglect, 44*, 159–69.

8. Temple, J. R., Choi, H. J., Brem, M., Wolford-Clevenger, C., Stuart, G. L., Peskin, M. F., & Elmquist, J. (2016). The temporal association between traditional and cyber dating abuse among adolescents. *Journal of Youth and Adolescence, 45*(2), 340–349.
9. Yahner, J., Dank, M., Zweig, J. M., & Lachman, P. (2015). The co-occurrence of physical and cyber dating violence and bullying among teens. *Journal of Interpersonal Violence, 30*(7), 1079–1089.
10. Zweig, J. M., Dank, M., Yahner, J., & Lachman, P. (2013). The rate of cyber dating abuse among teens and how it relates to other forms of teen dating violence. *Journal of Youth and Adolescence, 42*(7), 1063–1077.
11. Citron, D. K. (2009). Law's expressive value in combating cyber gender harassment. *Michigan Law Review*, 373–415.
12. Mantilla, K. (2015). *Gendertrolling: How misogyny went viral*. ABC-CLIO.
13. Bauman, S., Toomey, R. B., & Walker, J. L. (2013). Associations among bullying, cyberbullying, and suicide in high school students. *Journal of Adolescence, 36*(2), 341–350.
14. Jakobsson, M. (Ed.) (2016). *Understanding Social Engineering Based Scams* (pp. 103–113) https://doi.org/10.1007/978-1-4939-6457-4_10; and Rege, A. (2009). What's love got to do with it? Exploring online dating scams and identity fraud. *International Journal of Cyber Criminology, 3*(2). https://www.cybercrimejournal.com/aunshulregedec2009.htm
15. Shires, J. (2018). Enacting expertise: Ritual and risk in cybersecurity. *Politics and Governance, 6*(2), 31–40. https://doi.org/10.17645/pag.v6i2.1329

Chapter 3
Cybersecurity Analysis and Phishing Attack

Bhaswati Sahoo and Prasant Kumar Pattnaik

3.1 Introduction

Recent advances in embedded systems design, communication protocols, sensor technology, and mobile computing are enabling the development of a new class of system that integrates cyber space and our physical environment. While those working in the area of embedded system design are focused on designing computational models for specific applications, those working on these cyber-physical systems (CPS) are focused on establishing communications models that can reliably integrate time and feedback control into the model [1]. These CPS can assist us in monitoring and modifying the physical world in which we live and enhance our daily lives. Applications in aircraft and vehicle control systems, factory automation, weather forecasting, and deep sea drilling have already been identified [2].

The healthcare is undergoing a paradigm shift to integrate communication and information technologies with physical medical devices into a distributed network ensuring real-time and near real-time data transfer from the physical world to the cyber space for computation, data management, and analysis. In other words, the healthcare network can be seen as a cyberphysical system (CPS) defined in the literature as the emerging collection of distributed cyber systems that controls based on a set of rules, the interconnected physical systems networked by control loops to deliver a personalized medical assessment or even to predict medical events. It integrates sensing (monitoring of physical world parameters and variables), networking, computation technology, cognition by machine learning, and autonomy for the system to respond in real time without or with less human intervention.

B. Sahoo (✉) · P. K. Pattnaik
School of Computer Engineering, KIIT Deemed to be University, Bhubaneswar, India
e-mail: bhaswati.sahoofcs@kiit.ac.in; patnaikprasantfcs@kiit.ac.in

D. Mishra et al. (eds.), *Communication Technology and Gender Violence*, Signals and Communication Technology, https://doi.org/10.1007/978-3-031-45237-6_3

With the growing network of sensors and input devices capturing complex and variable physical data, new dependencies are introduced and new security vulnerabilities appear due to the fact that efficient methods for protecting data and securing communication networks are not yet developed to cover the continuous evolving healthcare network.

Although this is an emerging area and it is expected that it will be a global integration, there are some aspects to consider in the development of such a network from its early stages like security of data.

Healthcare data has significant value and is a potential target for hackers.1 2 Phishing is a method of attempting to gain potentially valuable details, such as usernames, passwords, or medical data, for malicious reasons, using targeted communications such as email or messaging in which the attacking party encourages recipients to click links to websites running malicious code or to download or install malware. Since phishing typically requires the recipient to perform an action, it relies on social engineering techniques, with many contacts therefore appearing to be from trusted sites such as financial institutions, or in the case of healthcare data, IT administrators or healthcare staff. Phishing refers to this general approach, in which large numbers of untargeted communications are sent to a wide range of recipients in the hope that a minority will become victims. Variants include spear phishing, in which communications are directed at specific individuals, or types of individuals or companies; clone phishing, in which a legitimate email has content changed to create a cloned email containing malicious content; and whaling, in which communications are targeted specifically at senior high-profile targets, often supposedly originating from "C-suite" or legal departments.3 4 The aim of this study is to report on an internal investigation into phishing targeting healthcare staff at one institution representing a UK National Health Service (NHS) hospital and review the medical peer-reviewed literature regarding phishing affecting healthcare organizations.

3.2 Objective

The objective of this study was to provide an overview of the literature at the intersection of cybersecurity and healthcare delivery. Health and medical centers use medical IoT devices for monitoring patients' treatment every day. Cybersecurity issue has to be the high rank for such systems because any technical mistake can cause harmful consequences, even a death. This paper represents analysis of cybersecurity-related standards and regulations, which are needed for building the trust IoT system based on international regulations for IoT, healthcare, medical devices, and Protected Healthcare Information (PHI) combined together. A detailed local cybersecurity audit was performed by our organization using a commissioned party along with standard penetration testing approaches as part of routine cybersecurity policy activity. Specific details of the methods and detailed findings of potential vulnerabilities are not provided for obvious reasons, but an overview of the strategy used is provided below. It is represented in the form of a hierarchical cybersecurity model, which is dedicated to the essential issue of healthcare security for each layer of healthcare and medical devices' architecture.

The healthcare network is based on the integration of information and communication technology to exchange data between the healthcare applications. Introducing a communication network in the healthcare systems leads to vulnerability dependencies, many inherited from the IT integration. Such vulnerabilities are based on:

- Disruptive attacks to shut down hospital systems, critical equipment, lab equipment, resetting configuration settings of medical devices (e.g., insulin infusion pumps) or rebooting life-sustaining devices;
- Identity theft and insurance fraud by impersonating and stealing patience records.
- Loss of historical medical information which is critical for treating patients with serious clinical illnesses.
- Access to controlled substances from laboratories [2, 3].

Although the integration of medical services with the information and communication technologies marks an important evolutionary step in taking the medical field to the next level, the security of these technologies can be seriously disturbed by the physical equipment of the health network and vice versa. Thus, the development of cybersecurity assessment technique which is based on the refined structure of the Advanced Security Assurance Case is needed.

In cybersecurity, a vulnerability is a weakness that can be exploited by cybercriminals to gain unauthorized access to a computer system. After exploiting a vulnerability, a cyberat-ack can run malicious code, install malware, and even steal sensitive data.

Vulnerabilities can be exploited by a variety of methods including SQL injection, buffer overflows, cross-site scripting (XSS), and open-source exploit kits that look for known vulnerabilities and security weaknesses in web applications.

Many vulnerabilities impact popular software, placing the many customers using the software at a heightened risk of a data breach, or supply chain attack. Such zero-day exploits are registered by MITRE as a Common Vulnerability Exposure (CVE).

3.3 Definitions of Vulnerability

There are many definitions of vulnerability. Here is a list of definitions from various network security authorities.

National Institute of Standards and Technology (NIST) Weakness in an information system, system security procedures, internal controls, or implementation that could be exploited or triggered by a threat source.

ISO 27005 A weakness of an asset or group of assets that can be exploited by one or more cyber threats where an asset is anything that has value to the organization, its business operations, and its continuity, including information resources that support the organization's mission.

IETF RFC 4949 A flaw or weakness in a system's design, implementation, or operation and management that could be exploited to violate the system's security policy.

ENISA The existence of a weakness, design, or implementation error that can lead to an unexpected, undesirable event compromising the security of the computer system, network, application, or protocol involved.

The Open Group The probability that threat capability exceeds the ability to resist the threat.

Factor Analysis of Information Risk The probability that an asset will be unable to resist the actions of a threat agent.

ISACA A weakness in design, implementation, operation, or internal control. Healthcare data has significant value and is a potential target for hackers.1 2 Phishing is a method of attempting to gain potentially valuable details, such as usernames, passwords, or medical data, for malicious reasons, using targeted communications such as email or messaging in which the attacking party encourages recipients to click links to websites running malicious code or to download or install malware. Since phishing typically requires the recipient to perform an action, it relies on social engineering techniques, with many contacts therefore appearing to be from trusted sites such as financial institutions, or in the case of healthcare data, IT administrators or healthcare staff.

Phishing refers to this general approach, in which large numbers of untargeted communications are sent to a wide range of recipients in the hope that a minority will become victims. Variants include spear phishing, in which communications are directed at specific individuals, or types of individuals or companies; clone phishing, in which a legitimate email has content changed to create a cloned email containing malicious content; and whaling, in which communications are targeted specifically at senior high-profile.

Unauthorized modification or injection of false data can impact medical decisions. This can be through the internet by bypassing firewalls or direct, using a portable device such as a USB stick introduced into the computer (e.g., within the hospital) to flood false records into the system. The methods vary and the attacker can use:

Network packet detection to make use of critical system information, like user account information and passwords, network topology. It can insert new information or change existing one in the packet [4].

Spoofing attack (IP spoofing) to impersonate an authorized user without having to use its password by falsifying data to gain access to the system [5].

Password attacks (like, password guessing, capturing, cracking – like dictionary, brute force, phishing attack – and sniffing) to gain access to accounts and services [4].

SQL injections to bypass authentication and tamper with database data (insert, update, and delete).

Deception attacks or even stealthy deception attacks are another form of false data injection [6] to introduce into the communication signal an incorrect sensor measurement or control input, an incorrect time stamp, or a wrong identity of the sending device.

Intrusion attacks (also, compromise the confidentiality objective) to illegally access the cyber system, e.g., malwares like backdoors.

Eavesdropping or sniffing – this is a passive attack in which the attacker listens and captures the network traffic packets.

Application layer attack – this kind of attack targets the application servers to cause a fault by reading, deleting, or even modifying data. It enables the attacker to introduce a packet sniffer into the internal network to gain confidential information or introduce viruses to spread and cause system failures.

The necessity of data being available is not as critical as confidentiality in a medical health network.

Although if a form of attack targets to disrupt the medical equipment in a hospital or clinic putting in danger people's life, then, availability becomes a primary concern. For example, denial-of-service (DoS) attacks target TCP/IP-based protocols (e.g., IEEE11073 – a group of communication standards for health devices with computer systems) altering the communication and determining incorrect decisions due to incomplete information or shut down critical hospital systems and even crash the system of all the testing equipment in a laboratory.

DoS types can be one of the following [7]:

Teardrop attack modifies the fragmentation in sequential IP packets. A machine that receives the fragmentation packets is unable to reassemble them, ending in device failure.

SYN attacks (synchronization request) exploit the three-way handshake of the TCP protocol. It overflows the limit space of the buffer with connection request, making the target system to ignore the legitimate requests.

Another attack is *Ping of Death* (in the buffer overflow category) where a large packet ping that exceeds the protocol specification limit is sent to the target system freezing it and exploiting the Internet Control Message Protocol (ICMP). The solution consists in rebooting, but in terms of a healthcare equipment, a freeze time is not acceptable because delaying command signals or interrupting real-time operations leads to physical damage and life-threatening conditions. *Smurf attack* is a brute force method that uses ping flooding and the destination address of the ICMP echo packet requests is the broadcast address of the network to congest the target network with traffic. Another type of attack that compromises the availability of data is the *jamming attack* that can have life-threatening effects because of the remote monitoring capability of the healthcare network.

There is a thin but obvious line between safety and security. *Security* refers to the protection of data and prevention of accessing or modifying it by unauthorized entities, while *safety* refers to people not being harmed because of false information, lack of it or information getting into the wrong hands. Furthermore, a damaged update to the medical software can change health records and information about medication [8]. Attacks may come from within the healthcare organization or from the outside targeting the communication network.

The inside attack can have the greatest impact of the overall system if internal security policies are loosely implemented or nonexistent, because the attacker has access to a wide variety of resources within the hospital, clinic, or research centers. These attacks can be unintentional (the use of network for nonbusiness purposes) and in this case the most frequent effect would be to infect the inside network with malware, or deliberate (access of restricted network services) which in this case can lead to serious damage because the attacker has a strong understanding of the overall system and can cover up easily its actions. Malware (malicious software) can target all the objectives containing viruses, Trojan horses, trapdoors, and backdoors. *Social engineering attacks* involve non-technical skills, like communication, persuasiveness, etc. Cyberattacks can occur at any connection of the network and at any endpoint. Protecting all the entry points is almost impossible but a better understanding of the possible attacks can help preventing and implementing the appropriate secure methods. For instance, a cyberattack has a series of stages before acting [9]: the targeted system is well researched; vulnerabilities of the targeted system are mapped to be exploited; the attacker runs a series of exploits on the entry points that are vulnerable. Usually an attacker can run a scanner to check whether a port is left open or to check for system vulnerabilities that were not patched. Also, the traffic data can be monitored and intercepted with a sniffer to capture login credentials. An attacker can trick the user into introducing its credentials on a fake site or use email phishing (very popular and effective because the email has legitimate information taken from other sources so as to make the victim trust the content). This attack is also known as man-in-middle attacks; once the attacker has access to the system it installs malware/viruses/rootkits and backdoors to establish a full control; having the system infiltrated the malware moves laterally to increase access and gain knowledge about the system; it collects information and possible transferring it via an internet connection, then destroying the system or deleting its confidential data, or just tempering with its operations. In a healthcare network these cyber-physical attacks targeting medical equipment can have serious consequences with immediate outcome. An overview of the possible effects of cyberattacks on such devices are described in the following section [10].

3.4 Conclusion

Hospitals receive a significant volume of potentially malicious emails. While many staff appear to be aware of phishing and respond appropriately, ongoing education is required across the spectrum of cybersecurity, with specific emphasis around "leakage" of information on social media. In order to make a determination of which architecture was the most appropriate from a security perspective candidate architectures were judged primarily on the number of security issues that had a DREAD ranking of important or higher. The current transition of the healthcare system resides on the integration of information and communication technologies with the physical equipment to offer an interconnected system for allowing real-time data to flow, better real-time monitoring of the patient's health status and improve overall quality in health services. In this paper are presented the most important issues that must be taken into consideration when designing the architecture of a secure, reliable healthcare network. It is not a solution to the problem, but an underlining of those elements that must be taken into account when the healthcare system is designed. We address the most important challenges associated with medical devices and data protection of the cyber-physical system, like the healthcare. Moreover, when developing the network's architecture, mechanisms for verifying the validity of data communicated, exchanged, and analyzed must be set in place. We stress out the fact that a security breach in the healthcare system can have a negative impact on the cyber aspect regarding medical data that may lead to incorrect health decisions and, also, on the physical aspect regarding medical devices that are connected to the patients and any failure can lead to life-threatening concerns.

References

1. Wright, A., Aaron, S., & Bates, D. W. (2016). The big phish: Cyberattacks against US healthcare systems. *Journal of General Internal Medicine, 31*, 1115–1118.
2. Jarrett, M. P. (2017). Cybersecurity—A serious patient care concern. *JAMA, 318*(14), 1319–1320.
3. Gudla, C., Rana, M. S., & Sung, A. H. (2018). Defense techniques against cyber attacks on unmanned aerial vehicles. In *Proceedings of the international conference on embedded systems, cyber-physical systems, and applications (ESCS)*. The Steering Committee of The World Congress in Computer Science, Computer Engineering and Applied Computing (WorldComp).
4. Anawar, S., Kunasegaran, D. L., Mas'ud, M. Z., Zakaria, N. A., et al. (2019). Analysis of phishing susceptibility in a workplace: A big-five personality perspectives. *Journal of Engineering Science and Technology, 14*(5), 2865–2882.
5. Sommestad, T., & Karlzén, H. (2019). A meta-analysis of field experiments on phishing susceptibility. In *2019 APWG symposium on electronic crime research (eCrime)*. IEEE.
6. Jalali, M. S., et al. (2019). Health care and cybersecurity: Bibliometric analysis of the literature. *Journal of Medical Internet Research, 21*(2), e12644.
7. Yaacoub, J.-P., et al. (2020). Security analysis of drones systems: Attacks, limitations, and recommendations. *Internet of Things, 11*, 100218.
8. Abdelhamid, M. (2020). The role of health concerns in phishing susceptibility: Survey design study. *Journal of Medical Internet Research, 22*(5), e18394.
9. Sethuraman, S. C., Vijayakumar, V., & Walczak, S. (2020). Cyber attacks on healthcare devices using unmanned aerial vehicles. *Journal of Medical Systems, 44*(1), 29.
10. Ly, B., & Ly, R. (2021). Cybersecurity in unmanned aerial vehicles (UAVs). *Journal of Cyber Security Technology, 5*(2), 120–137.

Chapter 4
Sentimental Analysis of Quarantine Fears Among Women Through NVIVO

Shivani Agarwal, Vijender Kumar Solanki, and Gloria Jeanette Rincón Aponte

4.1 Introduction

The world was suffering from Covid-19 pandemic [1]. Fear is usually *"experienced as an unpleasant emotion triggered by a constraining belief that something unwelcome is about to happen, or something is dangerous, or the safety of someone is threatened"* [2]. Surprisingly, the more we resist what is happening to us, the more fearful we become, and the more we suffer (https://www.ncbi.nlm.nih.gov/pmc/articles/PMC7435204/). Fear is of different parameters like fear of crowded place, fear of insects, fear of airplane, fear of water, and fear of loneliness. Among all the fears prevalent in the society, the fear of loneliness which is known as autophobia has predominantly encroached the women staying in quarantine.

The study discusses the rise of fear in the women lives in quarantine during Covid-19. The people dealing with the corona phobia need to pay a lot of attention as it creates a huge impact on the well-being of the person who is a part of quarantine. The term quarantine was defined as *"to stay away from all the people and live in a separate place if you have been in contact with any infected person of COVID-19"*. Corona virus was first identified in Wuhan, China 2019. Gradually it spread worldwide and acknowledged as pandemic by WHO. After a lot of research, scientist found a way to stop the growth of Covid-19 by isolating the infected person in home or at any place.

S. Agarwal (✉)
Department of HR, Galgotias University, Greater Noida, India, Ghaziabad

V. K. Solanki
Department of Computer Science and Engineering, CMR Institute of Technology, Hyderabad, India

G. J. R. Aponte
Facultad de Ingeniería, Universidad Cooperativa de Colombia, Bogotá, Colombia
e-mail: gloriaj.rincon@campusucc.edu.co

D. Mishra et al. (eds.), *Communication Technology and Gender Violence*, Signals and Communication Technology, https://doi.org/10.1007/978-3-031-45237-6_4

The term quarantine was first originally developed in 1377 for a thirty-day isolation period passed a law establishing a trentino [3]. Several people consider that quarantine represents *"an unwarranted diminution of personal liberty, whereas others see it as an integral aspect of communicable disease control"*. There are four principles that should be met to be considered as Ethics of Quarantine, namely, harm principle, least-restrictive-means, reciprocity, and transparency principle [4]. Quarantine helps the community to save the people from unnecessarily infecting others. So, it's the moral obligation of infected people to self-quarantine himself/herself to save others [5]. While some people thought of quarantine as social obligation, some were reluctant to stay in quarantine due to the fear of loneliness in quarantine during Covid-19, fear of the spread of disease in quarantine during Covid-19, the fear of performing household chores in quarantine during Covid-19, the fear of performing office work in quarantine during Covid-19, the fear of not getting proper medical assistance in quarantine during Covid-19, the fear of Speedy Recovery in quarantine during Covid-19.

Covid-19 has equal probability toward men or women getting infected. But the circumstances create a lot of differences in infection. Also, to save the people from infection, the strategy to save more people quarantine fear has different ways to creep in among genders. Men and women, both genders have different aspects of living in the society. Men are more quiet, busy in their own schedule, in short, we can say that they are less social as compared to women. Their level of responsibility is also very different in Indian Society. Working women have more chances of getting infection and spreading the same.

This study was certainly focused on women academicians as they were the frontliners to educate the nation. The level of infection among women academicians was at peak during the Covid-19 second wave. The suffering was arising due to infection. But it added more and more pain as their responsibility toward their students, the nature of the job is to give lectures continuously for 1–2 hours at a stretch and sometimes in a day they have four lectures. To take lectures it requires a lot of stamina and strong breath. But because of the infection, Covid-19 created a sense of breathlessness, cough, fever, weakness, etc.

With the physical problems, the fear of the spread of disease in quarantine during Covid-19, the fear of performing household chores in quarantine during Covid-19, the fear of performing office work in quarantine during Covid-19, the fear of not getting proper medical assistance in quarantine during Covid-19, the fear of speedy recovery in quarantine during Covid-19 adds more pain and problem among women. So, this study was able to analyze the below-mentioned objectives.

4.2 Objectives

The main objectives of the chapter are as follows:

1. To analyze the fear of loneliness among women in quarantine during Covid-19.
2. To analyze the fear of the spread of disease among women in quarantine during Covid-19.

3. To analyze the fear of performing household chores among women in quarantine during Covid-19.
4. To analyze the fear of performing office work among women in quarantine during Covid-19.
5. To analyze the fear of not getting proper medical assistance among women in quarantine during Covid-19.
6. To analyze the fear of speedy recovery among women in quarantine during Covid-19.

4.3 Literature Review

Lewandowski [6] mentioned that the Covid-19 pandemic has distinctive impact on individual life performance, across the world. Covid-19 has affected billions of people and upended the lives of most individuals toward the globe. Depressive symptoms were seen as a major outcome of Covid-19 [7].

Brooks et al. [8] mentioned that there are several benefits of quarantine to the world but it also created psychological damages among human race. Fear has been one of the highly recurrent psychological reactions in people during this coronavirus pandemic [9]. High levels of fear are associated with anxiety and depression [10]. Coronaphobia is an interpreter of psychological distress during the pandemic crisis [11].

Enea et al. [12] suggested that the three variables were examined in Covid-19 pandemic namely loneliness, preoccupation with God, and death obsession with fear of Covid-19. Results showed that after an average isolation time of almost 37 days, the perceived levels of loneliness significantly increased in the sample and created chaos in the brain of women.

Fear is a known (for centuries and in response to previous infectious outbreaks such as the plague), yet common response to infectious outbreaks and people react in many and individualized ways towards the perceived threat (Usher et.al. 2020). Quarantine enhances the fear among people of spreading the disease to others as well.

Working mommies, unable to cope with work, kids, and household chores, have left their jobs to react to the heightened need for care work at home. Mummies have a never closing thought that they are in quarantine but are unable to restrict themselves from caring for their loved ones [13]. Women have a distinctive place at home, and by quarantining and demanding other family members to stay at home, their duties and roles may vary, exposing them to new barriers. Yoosefi Lebni et al. [14] mentioned the individual problems, social problems, health problems, and family problems faced by women during quarantine.

Jain et al. [15] mentioned that quarantine has an influence on mental health during Covid-19 pandemic. The mental health deteriorated which is expected to have happened because of isolation, neglect, and loneliness. It has multi-dimensional impact on our physical, mental, social, and emotional well-being [16, 17]. Consistent experience was stated during earlier coronavirus pandemics, such as SARS and MERS pandemic. The influence of Covid-19 among the population is also substantial as it is apparent across the world. In order to meet the scenario, the responsibility of a psychiatrist is likely to be crucial equally in the short and long term [18].

During Covid-19 the shortage of medicine created problem in speedy recovery during quarantine. Recovery took a lot of time and energy for the women especially as they were more emotional and psychologically weak when it comes to health issues.

4.4 Results and Discussion

4.4.1 Research Methods

The study was conducted on a sample of Educators working in Higher Education Institutions specifically women during Covid-19. The sample consisted of women who were infected during Covid-19 and kept under self-quarantine. The data was collected through convenience sampling with the prior permission of Education Institute. Questionnaires were sent to 100 Education Institutes, out of which only **72** Education Institutes replied and received a data of ***50*** working women who were under self-quarantine. Data cleaning processes were conducted. Finally, 30 working women's qualitative data was analyzed with the help of tools such as Word Cloud Analysis, and sentiments analysis. The reliability of data was found to be 0.87 and the validity was checked with the help of expert in the related psychology domain. The demographic details are as follows in Table 4.1.

Table 4.1 Demographical details

Demographic (n = 42)	No. of respondents	Percentage (%)
Age (in years)		
24–40	22	73.33
41–57	8	26.67
Gender		
Female	30	100
Education		
Postgraduate	10	33.33
PhD	20	66.67
Impacted from Covid		
Nuclear family	25	83.33
Joint family	5	16.67

4.4.2 *Initial Assessments*

4.4.2.1 Data Cleaning

Initially, the data received was first put to cleaning process in which grammatical errors and spelling errors were corrected. It was done by setting the different checks for different procedures such as *meaningful sentences, unnecessary usage of conjunctions, and articles were removed* from the data. Then, grouping of data takes place to get better clarity on data for further analysis.

After the cleaning of the data the most prominent data which was extracted from the actual data under different heads was shown in Table 4.2.

4.4.3 *Correlation*

The different variables of quarantine fear are mentioned in Table 4.3 representing the correlation between loneliness fear, disease spread fear, performing household chores fear, performing office work, medical assistance fear, and recovery fear.

4.4.4 *Word Clouds Analysis*

The results from word clouds are shown as below:

1. *To analyze the fear of loneliness among women in quarantine during Covid-19.* The word cloud analysis shows the fear of loneliness in quarantine during Covid-19 which shows the fear of not being able to meet parents, neighbors, wife, husband, friends, kids/children, unable to attend parties, go and buy groceries from the market (Fig. 4.1).
2. *To analyze the fear of the spread of disease among women in quarantine during Covid-19.*
   The word cloud analysis portrays the fear of the spread of disease by Covid-19. Several studies have shown that fear of disease spread in quarantine has increased in women. For this study, the main fear of spreading the disease among their loved ones which are as: husband, parents, mother, brother, siblings and sisters (Fig. 4.2).
3. *To analyze the fear of performing household chores among women in quarantine during Covid-19.*
   The word cloud analysis shows fear of performing household chores such as preparation of food, home cleaning, utensil cleaning, clothes, arrangements for kids, health maintenance of family, and arrangement of groceries. The word cloud analysis clearly indicates the fear of household chores during quarantine in Covid-19 (Fig. 4.3).

Table 4.2 Data sheet for analysis

S.No.	Loneliness fear	Disease spread fear	Performing household chores fear	Performing office work	Medical assistance fear	Recovery fear
1	Unable to meet parents	Old parents	Food preparation	Lectures to be taken	Doctors	Weakness
2	Unable to meet wife	Mother	Clothes wash	Attending meetings	Nurses	Headache
3	Unable to meet husband	Husband	Utensil wash	NAAC/NBA/NIRF work	Beds	Vomit
4	Unable to meet children	Brother	Kids classes	IQAC work	Hospitals	Fever
5	Unable to meet friends	Sister	Kids homework	Alumni interaction	Oxygen	Stress
6	Unable to meet neighbors	Siblings	Kids health	Virtual industrial visit	Medicines	Depression
7	Unable to go out	Kids/children	Home cleaning	Virtual guest lecture	Oximeter	Anxiety
8	Unable to go to party	In-laws	Arrangement of groceries	Class coordination	N 95 mask	Restlessness
9	Unable to get groceries	Neighbor		Timetable	Sanitizer	Cold
10		Maid		Internship	Remdesivir	Cough
11				Placements		
12				Conferences/seminars/ workshops		

Table 4.3 Correlation among variables

S.No.	Variables	Mean	SD	1	2	3	4	5	6
1	Loneliness fear	52.19	6.69	1					
2	Disease spread fear	39.71	4.97	0.79**	1				
3	Performing household chores fear	25.95	4.88	0.75**	0.72**	1			
4	Performing office work	10.96	3.12	0.67**	0.64**	0.88**	1		
5	Medical assistance fear	32.15	5.04	0.79*	0.75*	0.77*	0.57*	1	
6	Recovery fear	15.95	3.99	0.88	0.66**	0.64*	0.82*	0.77*	1

Fig. 4.1 Fears of loneliness among women in quarantine. (Source: Authors own Work)

Fig. 4.2 Fear of the spread of disease among women in quarantine. (Source: Authors own Work)

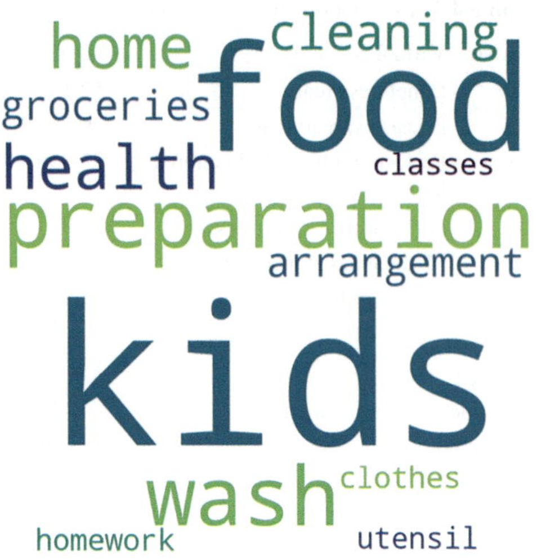

Fig. 4.3 Fear of performing household chores among women in quarantine. (Source: Authors own Work)

Fig. 4.4 Fear of performing office work among women in quarantine. (Source: Authors own Work)

4. *To analyze the fear of performing office work among women in quarantine during Covid-19.*
   The below word cloud analysis shows the fear of performing office work in quarantine. The women data belongs to academic profile, so the list of fears is to complete the work, lecture, NAAC, NIRF, Industrial Interaction, Alumni Meetings, Class Coordinator, Timetable, Workshops, Placements, etc. So, the cloud clearly mentions all the work attached with the educator's profile (Fig. 4.4).
5. *To analyze the fear of not getting proper medical assistance among women in quarantine during Covid-19.*
   The word cloud shows the fear of not getting proper medical assistance in quarantine such as availability of oxygen, hospitals, beds, nurses, doctors, mask,

Fig. 4.5 Fear of getting proper medical assistance among women in quarantine. (Source: Authors own Work)

Fig. 4.6 Fear of speedy recovery among women in quarantine. (Source: Authors own Work)

Oximeter, Sanitizers, Remdesivir, etc. This creates and exponentially enhances the fear among women when they feel and see the shortage of medical facilities worldwide (Fig. 4.5).

6. *To analyze the fear of speedy recovery among women in quarantine during Covid-19.*

   The word cloud analysis shows the recovery issues and the fear creep in among women which is weakness, anosmia, depression, headache, restlessness, cough, headache, depression, fever, cold, etc. The level of stress is enhanced as the recovery speed is very slow (Fig. 4.6).

4.4.5 Sentimental Analysis

Table 4.4 represents the women sentimental analysis which illustrates the fear while staying in quarantine during Covid-19. The women fear enhanced in quarantine as they were all alone staying in their room/or a separate place. Women were mentally engrossed with the fears of quarantine like how to handle the news flash continuously which shows the death, health issues, family, and peer pressure which are clearly identified in Table 4.4.

The first factors such as loneliness fear was having a very negative impact on the psychology of the women in the Indian context. Women were being afraid which is considered as a psychological loneliness fear had a moderately negative impact on respondent's sentiment. Women suffered from the fear of missing husband, missing their kids, missing their siblings, missing their parents, and missing their friends. On the other hand, women were advised to quit watching the news during Covid-19 quarantine time.

The second factor namely fear of disease spread also had huge negative impact on the women educators as they were scared for their loved ones including family and friends. Women educators were unable to take leaves from their workplace as they are the frontliners at the time of Covid-19. So, it is advised to them to take proper care of themselves not to meet anyone if in case they are infected. This is the way to save themselves and to their loved ones as well.

The third factor includes performing household chores which creates a moderately negative impact on women. In the Indian context women were fully engaged with the household chores. In fact, during the period of Covid-19, the maids were also not there to share the household chores. So, infected women who were under quarantine were feeling very negative of how to handle the infection, weakness, and all the household chores.

The fourth factor includes performing office work which created a huge negative impact on infected women. Working women were very much concerned and disturbed about the deadlines of the office work. Office work includes so many challenges such as taking lectures by an infected person creates breathlessness. The level of breathlessness creates restlessness. The work of NIRF, NBA, IQAC,

Table 4.4 Sentiment analysis codes

Codes	A: Very negative	B: Moderately negative	C: Moderately positive	D: Very positive
Loneliness fear	4	3	0	0
Disease spread fear	5	4	1	0
Performing household chores fear	3	3	2	0
Performing office work	4	4	1	0
Medical assistance fear	5	5	1	1
Recovery fear	1	2	4	5

Source: Authors own Work

timetable, class and timetable coordination, etc. creates a psychological burden on the infected person.

The fifth factor includes medical assistance fear because there was a huge shortage of medicines, vacant beds in hospitals, sanitizers, masks, and oximeter. All the emergency requirements created a negative impact on the infected women as they were scared of not getting the medical assistance in case, they required the same.

Lastly, the recovery fear was less among women as they were more positive that it shall pass, and they will be fit and healthy after this infection. But recovery takes time. The infection created a lot of weakness, depression, stress, and fear of losing their loved ones. The emotional impact on the health was positive and women were considered as a fighter for the same. But at a phase of recovery the family has a pivotal role to play during Covid-19.

4.4.6 Thematic Analysis

Table 4.5 reveals the results of the thematic analysis which consists of four main themes such as Quarantine, Women Educators, high moment in life, and low moment in life which has an essential part to play during Covid time. The loneliness fear, fear of spread of disease in quarantine during Covid-19, the fear of performing household chores in quarantine during Covid-19, the fear of performing office work in quarantine during Covid-19, the fear of not getting proper medical assistance in quarantine during Covid-19, and the fear of speedy recovery in quarantine during Covid-19 add more pain and problems among women. The major themes that emerge were *loneliness fear* and *fear of spread of disease*. So, to understand the theme and take proper precaution to reduce the fear among working women is of prime importance.

Table 4.5 Thematic analysis

Codes	Loneliness fear	Disease spread fear	Performing household chores fear	Performing office work	Medical assistance fear	Recovery fear
Quarantine	4	4	2	3	2	1
High moments in life	0	0	0	1	1	1
Low moments in life	4	4	1	1	4	1
Women educator	4	4	4	2	2	2

4.5 Conclusion

Quarantine helps in reducing the infection of Covid-19 among millions of people and saves the lives of human beings around the globe. The fear creeps in when people stay in quarantine. There is a huge list of fears among women which starts with loneliness fear, disease spread fear, pending household fear, pending office work fear, unavailability of medical facilities fears, and the impact of recovery fear on women race.

During the Covid wave, a lot of women lost their physical and mental health as they were not able to cope with the fears which were arising in their heads. So, the chapter shows the avenues for the policymakers of the health department to create new poliçies for quarantine. People know the basic rules of quarantine. But what they are lacking is how to handle those fears. The policy should consist of solutions so that the infected person should not feel lonely. He/she should be well versed with the love, and affection from their family and friends. Office/bosses should also understand the fear among the infected person and show empathy toward the infected one. The government should provide proper medical assistance for the person and the family of the infected person. Also, the recovery of the person should be of prime importance. The recovery policy consists of love, entertainment, and counseling of the infected person.

4.6 Future Work

The study suggests the list of fears that creep in while women in self-quarantine. The Terror Management Theory (TMT; [19]) suggests the future work for other researchers which can be considered to find the solutions for the fear of the women in quarantine. The future study can also check the impact of quarantine on other variables such as mental well-being, subjective well-being, level of trust [20], etc.

References

1. Agarwal, S., Tyagi, M., & Bhardwaj, A. (2022). Sentimental analysis of fears, psychological disorders and health issues through NVIVO during second wave of Covid-19. In *Predictive analytics of psychological disorders in healthcare* (pp. 223–237). Springer.
2. Bell, J. M., Moules, N. J., & Wright, L. M. (2009). Therapeutic letters and the family nursing unit: A legacy of advanced nursing practice. *Journal of Family Nursing, 15*(1), 6–30.
3. Mackowiak, P. A., & Sehdev, P. S. (2002). The origin of quarantine. *Clinical Infectious Diseases, 35*(9), 1071–1072.
4. Upshur, R. (2003). The ethics of quarantine. *AMA Journal of Ethics, 5*(11), 393–395.
5. Harris, J., & Holm, S. (1995). Is there a moral obligation not to infect others? *BMJ, 311*, 1215–1217.

6. Lewandowski, S. L., Cardone, R. L., Foster, H. R., Ho, T., Potapenko, E., Poudel, C., et al. (2020). Pyruvate kinase controls signal strength in the insulin secretory pathway. *Cell Metabolism, 32*(5), 736–750.
7. Aknin, L. B., De Neve, J. E., Dunn, E. W., Fancourt, D. E., Goldberg, E., Helliwell, J. F., et al. (2021). Mental health during the first year of the COVID-19 pandemic: A review and recommendations for moving forward. *Perspectives on Psychological Science, 17*(4), 915–936. https://doi.org/10.1177/17456916211029964
8. Brooks, S. K., Webster, R. K., Smith, L. E., et al. (2020). The psychological impact of quarantine and how to reduce it: Rapid review of the evidence. *Lancet, 395*, 912–920. https://doi.org/10.1016/S0140-6736(20)30460-8
9. Wang, C., Pan, R., Wan, X., Tan, Y., Xu, L., Ho, C. S., et al. (2020). Immediate psychological responses and associated factors during the initial stage of the 2019 coronavirus disease (COVID-19) epidemic among the general population in China. *International Journal of Environmental Research and Public Health, 17*, 1729. https://doi.org/10.3390/ijerph17051729
10. Ahorsu, D. K., Lin, C. Y., Imani, V., Safari, M., Griffiths, M. D., & Pakpour, A. H. (2020). The fear of COVID-19 scale: Development and initial validation. *International Journal of Mental Health and Addictions*. https://doi.org/10.1007/s11469-020-00270-8
11. Lee, S. A., Jobe, M. C., Mathis, A. A., & Gibbons, J. A. (2020). Incremental validity of coronaphobia: Coronavirus anxiety explains depression, generalized anxiety, and death anxiety. *Journal of Anxiety Disorders, 74*, 102268. https://doi.org/10.1016/j.janxdis.2020.102268
12. Enea, V., Eisenbeck, N., Petrescu, T. C., & Carreno, D. F. (2021). Perceived impact of quarantine on loneliness, death obsession, and preoccupation with God: Predictors of increased fear of COVID-19. *Frontiers in Psychology, 12*, Article ID: 643977.
13. Moraes, C., Santos, J., & Assis, M. P. (2020). "We are in quarantine but caring does not stop": Mutual aid as radical care in Brazil. *Feminist Studies, 46*(3), 639–652.
14. Yoosefi Lebni, J., Abbas, J., Moradi, F., Salahshoor, M. R., Chaboksavar, F., Irandoost, S. F., ... & Ziapour, A. (2021). How the COVID-19 pandemic effected economic, social, political, and cultural factors: A lesson from Iran. *International Journal of Social Psychiatry, 67*(3), 298–300.
15. Jain, A., Bodicherla, K. P., Raza, Q., & Sahu, K. K. (2020). Impact on mental health by "living in isolation and quarantine" during COVID-19 pandemic. *Journal of Family Medicine and Primary Care, 9*(10), 5415.
16. Bäuerle, A., Teufel, M., Musche, V., Weismüller, B., Kohler, H., Hetkamp, M., et al. (2020). Increased generalized anxiety, depression and distress during the COVID-19 pandemic: A cross-sectional study in Germany. *Journal of Public Health (Oxford, England)*. https://doi.org/10.1093/pubmed/fdaa106
17. Kontoangelos, K., Economou, M., & Papageorgiou, C. (2020). Mental health effects of COVID-19 pandemia: A review of clinical and psychological traits. *Psychiatry Investigation, 17*, 491–505.
18. Banerjee, D. (2020). The COVID-19 outbreak: Crucial role the psychiatrists can play. *Asian Journal of Psychiatry, 50*, 102014. https://doi.org/10.1016/j.ajp.2020.102014
19. Greenberg, J., Pyszczynski, T., & Solomon, S. (1986). The causes and consequences of a need for self-esteem: A terror management theory. In *Public self and private self* (pp. 189–212). Springer.
20. Agarwal, S. (2020). Trust as a missing link between quality of work life and subjective well-being. *Ingeniería Solidaria, 16*(1), 1–21.

Chapter 5
Leveraging OSINT and Artificial Intelligence, Machine Learning to Identify and Protect Vulnerable Sections of Society

Smriti Panda and Oorja Rungta

5.1 Introduction

With most of human communication moving online, some unwanted aspects of human interactions have also found their way to online platforms namely bullying and harassment. Bullying has taken the form of cyberbullying. Cyberbullying or harassment is an attempt to make someone feel threatened, uncomfortable, or mocked thus undermining their self-esteem [1]. It is often detected by looking for signs of profane words, punctuation marks, personal pronouns, and emotion text in online interactions [2]. Traditionally bullying has signified some kind of physical or psychological harm that might be received from one's peers. With the advent of modern technology cyberbullying has emerged as a new phenomenon wherein bullying has moved to digital platforms full of young and naive victims [3]. Cyberbullying targets vulnerable populations like adolescents and women by inflicting mental or emotional abuse using digital platforms [4]. Cyberbullying often takes the form of psychological abuse in the form of offensive content which is hard to remove and track as it is often posted anonymously. According to a 2013 survey by the National Center for Educational Statistics (NCES), almost 30% of the students reported that they had been the victims of cyberbullying [5]. Modern tools allow for more methods that can be used to hurl abuse including but not limited to rumormongering, morphing pictures, or posting insulting comments on the unfortunate victim. These methods can be used on both public forums or via private messaging [3]. This is an unfortunate side effect of a rapidly developing virtual world, a problem that must be stemmed with great haste and efficiency. Fortunately, the solution to this problem posed by technology lies in technology itself. Recently a lot of research has

S. Panda (✉) · O. Rungta
School of Computer Science and Engineering, School of Cyber Security and Digital Forensics, VIT Bhopal University, Bhopal, Madhya Pradesh, India

D. Mishra et al. (eds.), *Communication Technology and Gender Violence*, Signals and Communication Technology, https://doi.org/10.1007/978-3-031-45237-6_5

been undertaken to use artificial intelligence and machine learning (AI&ML) sometimes on its own and sometimes in conjunction with Open-Source Intelligence (OSINT) to prevent and detect cyberbullying and harassment. In some cases this research has been extended to cover post-incident therapy to help a survivor of such traumatic incident recover and cope with latent issues generated from the incident in question.

OSINT is the process of accumulating and analyzing data available publicly from open sources such as print newspapers, blogs, social networks, public government, reports, or professional and academic publications. This is a massive influx of information which is very hard to make sense of. Technology provides us tools to gather and also analyze this data for effective usage [3]. Artificial intelligence is a subdomain of Computer Science that aims to create intelligent systems that can imitate natural phenomenon of higher order thinking like speech, comprehension, visual perception, learning, reasoning, planning, etc. [6]. Machine learning aims to create system that can learn on their own. Cyberbullying and harassment are subjects where context, sentiment, and intent analysis are important. These subject matters are often studied under cyberpsychology and are difficult to define in terms a machine can understand. Nevertheless with progressive breakthroughs in AI and ML that are making machines more humanlike taking context, sentiment, and intent into account in an attempt to automate and detect cyberbullying and harassment is increasingly becoming a viable option.

Training AI and ML models need a dataset. Considering the problem at hand that is cyberbullying and harassment on online platforms directly getting data to train such models from these platforms seems the most organic and efficient method. This is where OSINT comes into play. OSINT tools allow researchers to gather and process massive amounts of open-source data. This processed data can then be used to conduct further studies. In this particular instance, the data gathered by OSINT will be used to train the AI and ML models.

5.2 Detailed Description

Social media platforms have allowed for communication across borders in real-time and these platforms have seen massive amounts of traffic. For instance, Twitter alone sees more than 500 million tweets a day [7]. This shows how large a forum of communication social media platforms have become. Yet they need to be regulated to prevent untoward activities like cyberbullying, harassment, and cyberstalking. Previously researchers have used Naive Bayes algorithm, K-Nearest Neighbors algorithm, Logistic Regression algorithm, Decision Tree algorithm, Random Forest algorithm, Linear Support Vector Classifier, Adaptive Boosting algorithm, Stochastic Gradient Descent algorithm, and Bagging classifiers to detect cyberbullying on twitter datasets [7]. Researchers have also used deep learning with neural networks like CNN, hybrid CNN-LSTM, and mixed CNN-LSTM-DNN to perform language analysis on social media posts taken from Google News, Twitter, or Formspring.

They used pre-defined words, a Bag-of-Words, and semantic features to detect cyberbullying with an SVM classifier [3]. Usage of Natural Language Processing (NLP) is another common and highly researched method to detect cyberbullying, harassment, and cyberstalking. As discussed above cyberbullying, harassment, and cyberstalking are subject matters of the cyberpsychology domain whose detection requires an understanding of context. NLP is a good tool to perform such an analysis. It is a sub-domain of AI and ML that is responsible for establishing a communication bridge between humans and machines [3]. It allows a machine to navigate and comprehend the ambiguous world of natural language that is context-dependent and full of linguistic devices that change the literal meaning of something [3]. It is currently being utilized to create chatbots and perform speech recognition [3]. Recent research also shows its effectiveness in detecting cyberbullying, harassment, and cyberstalking.

Most research into cyberbullying, harassment, and cyberstalking detection needs certain markers to determine whether a piece of text is intended to make someone feel bullied or harassed. It is these markers that the AI and ML algorithms look for when trying to detect cyberbullying, harassment, and cyberstalking [8]. These markers include textual features like explicit words, punctuation marks, uppercase, personal pronouns, emotion text, hashtags, and URLs or Network features such as the number of account followers and following, number of likes, number of shared media and mentioned users, date of creation of the account and various features exclusive to the user [2]. Some researchers hire people to annotate data, that is manually go through material and classify it as bullying or not bullying. This annotated data is then used to train AI and ML models. But manual annotation introduces human bias which can reduce the quality of the dataset. The number of publicly available dataset to detect cyberbullying are limited [1].

Modern research conducted in this domain suggests using AI and ML and OSINT in conjunction with each other. OSINT is a massive source of data and will allow researchers to directly get data on cyberbullying, harassment, and cyberstalking from the source – social media. Collecting data from OSINT sources to train AI and ML models will also reduce human bias and lead to higher quality datasets. Although OSINT data comes with its own challenges due to a lack of standard format. Nevertheless, solutions are being created to improve and process the data gathered from OSINT as its usage in AI and ML and big data analytics increases [9]. One excellent example of technology that uses OSINT and AI and ML cohesively to detect harassment on online platforms is CyberDect [3]. It combines NLP with OSINT to analyze posts with profane words and abuse toward the victim [3]. CyberDect uses OSINT to gather data from the victim's profile and his followers. The gathered data is analyzed using NLP to generate a report which shows things from both the stalker and victim profile helping analyze possible cases of cyberbullying [3].

5.3 Applications of This Technology

The practice of using of open-source intelligence has been in place for decades in government agencies and law enforcement. It further emerged as a technique used in fields of business and academic research by analyzing trends in publicly accessible data. This has led to the creation of information-collecting tools that made the traditional method of manual information harvesting easier and more productive. Current applications of OSINT involve other prominent technologies such as crowdsourcing, data mining, machine learning, artificial intelligence, etc., to tackle a variety of issues pertaining to military, cybersecurity, and several other areas. AI and ML have been in the research domain for a while but are still being used to make inroads in various domains and are very much relevant to the problem being discussed in this paper. The following section analyzes the applications of both OSINT and AI and ML in detecting cyberbullying harassment and cyberstalking.

Antonio L'opez-Mart'ınez et al. have created CyberDect an online tool that combines OSINT with NLP to analyze posts on social media for abusive language and profanity toward the victim and compiles it into a report that shows results from both the victim and perpetrator profile. It was tested using two real high school accounts from Spain and works with the Spanish language [3]. Choong Hon and Dewi Varanthan from the University of Malaya created a detection model that is used in twitter to detect cyberbullying. This system looks for negative keywords and profanity to determine whether a tweet is harmful and generates an alert once a harmful keyword is detected [10]. Nureni Ayofe Azeez et al. have worked on identifying and detecting cyberbullying on twitter using behavioral markers and machine learning together. Their experiments show that Random Forest Classifier has been the best at detecting cyberbullying with medians of 0.77,0.73 and 0.94 across the datasets. Despite their high-performing approach it has certain limitations. Twitter doesn't make age and gender information of users available. Their study is not inclusive to any other languages but English [7]. Gutiérrez-Esparza et al. used Random Forest classifier algorithm, OneR, and Variable Importance Measures (VIMs) to research cyber-aggression and classify it for Spanish language users. The classification in the above system divides aggressive posts into the following subcategories: sexual orientation-based violence and aggression, violence based on gender difference, and racial discrimination [11]. Nadhini et al. used Naive Bayes classifier and Lavenshtein calculation to arrange cyberbullying movement and detect it. They created a framework to classify cyberbullying into sub-categories of provocation, blazing, bigotry, and psychological oppression. The results in their framework exhibited a mean exactness from fornmspring.me as 93.79% and a precision of 94.59% from myspace.com [12]. Gomez-Adorno et al. worked primarily with the Spanish language to detect aggressive language. They grouped tweets into forceful and non-forceful and trained an ML model with it. They discovered 64.58% of non-forceful tweets and 35.42% of forceful tweets with their trained model [13].

Daniel automated detection of cyberbullying and harassment on YouTube using Support vector Machine Model and N-grams with a precision of 81.8%. He later improved it to 83.9% [14]. Haidar et al. diversified the detection of cyberbullying on Twitter and Facebook datasets to Arabic. They built a multilingual framework that used Naive Bayes classifier. Their research showed 33% perception of cyberbullying in Arabic [15]. Tosin et al. developed an ML model of Multinomial Naive Bayes and optimized linear Support Vector Machine (SVM) to recognize and intercept bullying messages. They achieved a 92% accuracy in detection. They also developed a web application programming interface (API) with flask python framework to make it easily accessible. Their API can be integrated with any social media website [16].

Mary Ann et al. have created a tool TwitterOSINT to extract and visualize data that is publicly available on Twitter. TwitterOSINT allows near real-time analysis of English-based Twitter posts which is then processed using NLP. They tracked several relevant keywords related to cyberpsychology to check whether certain concepts are trending or not to answer research questions [9]. This concept can also be applied to detect cyberbullying and harassment.

5.4 Possible Solutions

The necessity of reducing cyberthreats against women has led to numerous researches being conducted in order to propose solutions that, if integrated, might help to counter the rise of gender violence existing in cyberspace. This is a curated overview of some of these methods that are relevant in contributing to the cause to some extent by focusing on handling particular cybercrimes that might lead to general harassment of the victim. Out of these, one of the most widely known crimes is cyberstalking.

Cyberstalking in brief definition can be compared to traditional stalking that evokes negative emotions, gravely harming mental health that further results in fear of being victimized. This happens due to the intrusive communication patterns used by these stalkers toward the victims [17]. Though the conceptualization of cyberstalking is debated, its impact remains hard to ignore with victims experiencing psychosomatic and psycho-social problems as well as undergoing feelings of aggression, sleep disturbances, etc. [18]. Cyberstalking is majorly targeted at women which further adds up to the gender-based harassment online. To help this case, the proposed solution involves OSINT which makes use of username itself to identify victims and attackers. The algorithm goes through the process of user analysis through webscraping to collect data pertaining to social media such as followers, friends, follower count, etc. and analyses suspicious activity with the help of sentiment analysis. The attacker can be classified as active or a passive stalker who can be distinguished by the algorithm formed. The algorithm keeps scores of behaviors studied by applying the concept of NLP and discards unnecessary textual data using machine learning and AI models [19]. This automated process that detects the attackers of cyberstalking victims and identifies probable stalkers with its few drawbacks has a chance to be fruitful if improved and integrated and has a potential to greatly impact the investigation procedures of cybercrimes.

The advent of the internet has led to more and more engagement of the public that includes all ages, even children. The young of our society are oblivious to the threats persisting online and often get involved in unfortunate situations that could lead to child abuse that results in several damaging consequences. The popularity of social media and the young generation's involvement in it encourages the criminals lurking online to exploit them in harmful ways. Thus, to protect these vulnerable sections of society, one of the solutions researched uses crowdsourcing and the open-source intelligence to promote efficient investigation in online child abuse cases which often are way too complicated to be brought to justice quickly.

The theoretical model aims at solving the piling backlog of child abuse cases by helping the department of digital forensics. In the model, there is an active participation by volunteers among public that could serve as a pool of information which, if pieced together, could help in facilitating progress in investigations. This data and information being not completely reliable will undergo OSINT procedures and form reports which would be further analyzed by law enforcement agency for identifying relevant clues and discarding useless information. This process with proper digital examination and suitable analysis of information sets has the potential to reveal digital belongings of a suspected abuser which could lead to exact detection of locations and disclosure of real identities of criminal associates or child victims [20]. This model could be further developed to improve accuracy by involving AI methodology that could automatically sort important data so as to decrease much of manual involvement. This would not only reduce the workload but also would prove to be accessible in real time, establishing connections between threats and clues.

In terms of cybersecurity, the OSINT method which is currently in practice is used for criminal profiling by law enforcement agencies [21]. The open sources are used to protect citizens against cybercrimes by making them aware of potential threats, building security walls, warning about attacks, and educating them about precautions. The already applied procedures are crucial in securing the discussed members of society from becoming victims of these threats.

5.5 Prospect of These Models

If properly implemented, the above models would possibly turn out to be successful means to combat causation of various cyberthreats that are a part of gender violence in the online world and the attacks to other vulnerable portions of the community. The idea of correlating these technologies has plenty of advantages such as AI wearables that could provide open-source information instantly or AI ML models instilled in criminal investigation frameworks that boost accuracy and overall performance thus assisting in delivering justice efficiently and within less amount of time.

Openly available sources have their disadvantages. OSINT mechanisms are prone to misinformation as the information-gathering sources are mainly social media or other publicly accessible sources that are a hub of subjective opinions and lack of clarification [22]. In this case, while working with OSINT, we should take other techniques into account such as automated models that help in sorting accurate information.

The resulting methods might be readily available for usage as OSINT combined with artificial intelligence has scopes that continue to progress. In the areas of cybercrime research as well AI has proven to be useful in processing the volumes of information that involves posts, pictures, videos, etc. [23] that in turn is mostly harvested through open-source intelligence techniques.

5.6 Conclusion

The increasing scope of gender violence is threatening. In India alone, data from years 2017 to 2021 from the National Crime Records Bureau reveal growth in cases of cyber pornography, Hosting, and other forms of gender-based cybercrimes [24]. Cyberbullying comprising sexual cyberbullying, harassment, etc. has been proved to be prevalent in online community as apparent in multiple surveys [25]. Thus, the need for reliable systems to counter issues focused on but not limited to women must be handled with utmost urgency.

Open-source intelligence incorporated along with artificial intelligence holds the prospect of dynamic improvement in the field of cybersecurity. With successful executions in fields of cyber-defense, organized crimes, business, and criminal intelligence investigations, this technology has resolved dangers and issues lasting within the web. To push this methodology further and develop other potential models would serve as a boon to the effort of protecting our society. Two such models are discussed in this paper that have a scope of improvement and would be beneficial for future development to secure the walls of security.

References

1. Eronen, J., Ptaszynski, M., Masui, F., Leliwa, G., Wroczynski, M., Piech, M., & Smywiński-Pohl, A. (2022). *Initial study into application of feature density and linguistically-backed embedding to improve machine learning-based cyberbullying detection*. https://doi.org/10.48550/arXiv.2206.01889
2. Thun, L. J., Teh, P., & Cheng, C.-B. (2021). CyberAid: Are your children safe from cyberbullying? *Journal of King Saud University – Computer and Information Sciences, 34*. https://doi.org/10.1016/j.jksuci.2021.03.001
3. López-Martínez, A., et al. (2019). CyberDect. A novel approach for cyberbullying detection on twitter. In *International Conference on Technologies and Innovation*. Springer.

4. Gomez, C. E., Sztainberg, M. O., & Trana, R. E. (2022). Curating cyberbullying datasets: A human-AI collaborative approach. *International Journal of Bullying Prevention, 4*(1), 35–46.
5. Deliri, S., & Albanese, M. (2015). Security and privacy issues in social networks. In *Data management in pervasive systems* (pp. 195–209). Springer.
6. Evangelista, J. R. G., et al. (2021). Systematic literature review to investigate the application of open source intelligence (osint) with artificial intelligence. *Journal of Applied Security Research, 16*(3), 345–369.
7. Azeez, N. A., et al. (2021). Cyberbullying detection in social networks: Artificial intelligence approach. *Journal of Cyber Security and Mobility, 10*(4), 745–774.
8. Kaur, P., Dhir, A., Tandon, A., Alzeiby, E. A., & Abohassan, A. A. (2021). A systematic literature review on cyberstalking. An analysis of past achievements and future promises. *Technological Forecasting and Social Change, 163*, 120426. https://doi.org/10.1016/j.techfore.2020.120426. ISSN 0040–1625. https://www.sciencedirect.com/science/article/pii/S004016252031252X
9. Hoppa, M. A., et al. (2019). Twitterosint: Automated open source intelligence collection, analysis & visualization tool. *Annual Review of Cybertherapy and Telemedicine, 2019*, 121.
10. Hon, L., & Varathan, K. (2015). Cyberbullying detection system on twitter. *IJABM, 1*(1), 1–11.
11. Gutiérrez-Esparza, G. O., Vallejo-Allende, M., & Hernández-Torruco, J. (2019). Classification of cyber-aggression cases applying machine learning. *Applied Sciences, 9*(9), 1828.
12. Nandhini, B. S., & Sheeba, J. I. (2015). Cyberbullying detection and classification using information retrieval algorithm. In *ICARCSET '15*.
13. Gómez-Adorno, H., et al. (2018). *A machine learning approach for detecting aggressive tweets in Spanish*. IberEval@ SEPLN.
14. Ducharme, D. N. (2017). *Machine learning for the automated identification of cyberbullying and cyberharassment*. University of Rhode Island.
15. Haidar, B., Chamoun, M., & Serhrouchni, A. (2017). A multilingual system for cyberbullying detection: Arabic content detection using machine learning. *Advances in Science, Technology and Engineering Systems Journal, 2*(6), 275–284.
16. Ige, T., & Adewale, S. (2022). AI powered anti-cyber bullying system using machine learning algorithm of multinomial Naïve Bayes and optimized linear support vector machine. *International Journal of Advanced Computer Science and Applications (IJACSA), 13*(5), 5. https://doi.org/10.14569/IJACSA.2022.0130502
17. Dreßing, H., Bailer, J., Anders, A., Wagner, H., & Gallas, C. (2014). Cyberstalking in a large sample of social network users: Prevalence, characteristics, and impact upon victims. *Cyberpsychology, Behavior, and Social Networking, 17*(2), 61–67. https://doi.org/10.1089/cyber.2012.0231
18. Dughyala, N., Potluri, S., Sumesh, K. J., & Pavithran, V. (2021). Automating the detection of cyberstalking. In *2021 second International Conference on Electronics and Sustainable Communication Systems (ICESC)*, pp. 887–892. https://doi.org/10.1109/ICESC51422.2021.9532858
19. Açar, K.. (2017). OSINT by crowdsourcing: A theoretical model for online child abuse investigations. https://doi.org/10.13140/RG.2.2.11891.84004.
20. Hwang, Y.-W., Lee, I.-Y., Kim, H., Lee, H., & Kim, D. (2022). Current status and security trend of OSINT. *Wireless Communications and Mobile Computing, 2022*, Article ID 1290129., 14 pages. https://doi.org/10.1155/2022/1290129
21. Pastor-Galindo, J., Nespoli, P., Gómez Mármol, F., & Martínez Pérez, G. (2020). The not yet exploited goldmine of OSINT: Opportunities, open challenges and future trends. *IEEE Access, 8*, 10282–10304. https://doi.org/10.1109/ACCESS.2020.2965257
22. Madiwale, R., & Kumar, S. (2022). An outline on increasing online gender violence against women in India and the role of cyber security. *Journal of Positive School Psychology, 6*(6), 3920.
23. Pasha, S. A., Ali, S. & Jeljeli, R. (2022). *Artificial intelligence implementation to counteract Cybercrimes against children in Pakistan*. Hu Arenas. https://doi.org/10.1007/s42087-022-00312-8

24. Ehman, A. C., & Gross, A. M. (2019). Sexual cyberbullying: Review, critique, & future directions. *Aggression and Violent Behavior, Volume, 44*, 80–87. https://doi.org/10.1016/j.avb.2018.11.001. ISSN 1359-1789, https://www.sciencedirect.com/science/article/pii/S135917891830168X
25. Gundur, R. V., Berry, M., & Taodang, D. (2021). Using digital open source and crowdsourced data in studies of deviance and crime. In A. Lavorgna & T. J. Holt (Eds.), *Researching cyber-crimes*. Palgrave Macmillan. https://doi.org/10.1007/978-3-030-74837-1_8

Chapter 6
Cyber Risk and Gender Violence in Fashion Advertising

Sarita Tripathy

6.1 Introduction

In any gleaming high fashion magazine today you will be able to encounter images of women who have been exploited, sexually and mentally assaulted or brutalized and even murdered. The fashion world is shown in advertisements in the light of sex and violence [1] which is very bad for business and is directly or indirectly causing damage by making the violence against women a normal practice. By the various kinds of gestures like #MeToo and #TimesUp, different organizations have taken to social media to highlight the global impact of disproportionate violence. The term "Fast fashion" means "an approach to the design, create, and to marketing clothing fashions that exemplifies making very fast fashion trends and make them cheaply available to the consumers." It is the place where specifically women from all over the globe come.

The factories manufacturing clothing for companies of fast fashion have been criticized since many years, reason being their unsafe and inhumane working conditions. The women employees in these organizations are put through extremely strident work environments with scanty pay. In many countries like Bangladesh, Cambodia, Indonesia, India, and Sri Lanka the production companies have been under fire for putting workers to such type of conditions. Some specific quota of production are given to women in these companies which if not met with then they face grave consequences. A number of women have also shared their horrific experience of being beaten by the bosses for not reaching the target of making enough articles of clothing within the assigned time frame, some have also shared their experience of working for long hours with very less salary. These reports are countless.

S. Tripathy (✉)
School of Computer Engineering, KIIT Deemed to be University,
Bhubaneshwar, Odisha, India
e-mail: sarita.tripathyfcs@kiit.ac.in

D. Mishra et al. (eds.), *Communication Technology and Gender Violence*, Signals and Communication Technology, https://doi.org/10.1007/978-3-031-45237-6_6

Many Organizations like Global Labor Justice, Asia Floor Wage Alliance, and CENTRAL Cambodia are pushed for investigations of gender-based violence in major fast fashion companies in the hope that they will be forced to take action. A survey has been conducted for meeting such women who are victims of these poor labor regulations, and these organizations also highlight their demands: ending of violent conditions, adequate wages, and reasonable working hours.

Many multi-million dollar companies are blamed for subjecting models to inhumane conditions and sexual violence. It is very important that in fast fashion individuals [2] consume responsibly, as it is mostly accepted and used apparel options in the modern world. Education and awareness in this field is very much required to mitigate the harm the fast fashion companies are causing.

6.1.1 Need of Cyber Security in Fashion Industry

The fashion companies are now facing high risks like cyber attacks and breaches of data, by their customers and the wider economy. Years of hard work will be totally reverted by the theft of corporate, customer, and employee data or funds it will undermine relationships and have a significant impact on reputation and performance.

The attackers are able to have illegal access to the system through these attacks and many of cyber incidents [3] for the purpose of causing damage including attempts to gain illegal access to a system, network, infrastructure, or device. It has been found that the US data breaches were up 38% in the second quarter of 2021 compared with the first quarter, and breaches in the first half of the year alone reached 76% of the total reported in 2020.

Across the range of different fashion industry processes cyber security risks exist, from digital design and data analytics to online transactions and supply chain operations. There is recent digitization in many back-office systems, meaning they present a potential point of weakness for fashion leaders and security teams who had not previously been required to identify, assess, and mitigate potential risks in those areas. Indeed, shifting ways of working create constant challenges, requiring flexible decision-making muscle and continuous re-invention of defenses. There are two recent evolutions in fashion industry practices that have increased cyber vulnerabilities.

The first is a movement toward more agile ways of working. New products and services are increasingly developed and brought to market through fast-paced iterations using agile methods, where rapid timelines often do not allow for rigorous risk checks. Security teams must be involved early in the development process and embedded into the full digital life cycles of new products and services. The second is the ongoing evolution of technologies. Increasing use of cloud computing, artificial intelligence, and machine learning is exposing companies to more cyber risks by widening the scope for attack. Security teams must be innovative in finding ways to apply common security patterns and methods to new technologies.

Cyber risk is on a long-term upward trend that accelerated during the Covid-19 pandemic, partially as a result of widespread adoption of work-from-home patterns and technologies and soaring demand for e-commerce. Indeed, online retail has been one of the most attacked sectors over the past year, accounting for 10.2% of all attacks across industries. Given the growing frequency and severity of incidents, regulators are requiring businesses to protect themselves, their partners and their customers, and punishing those that fail to do so. Europe's General Data Protection Regulation (GDPR) imposes fines for non-compliance of as much as 4% of a company's global annual revenue.

A challenge for companies looking to invest in cyber defenses is that the cost of initiating an attack is significantly lower than that of protection. This creates an asymmetric battlefield in which hackers, companies, state-sponsored agencies, and other perpetrators can enter systems with relative ease. Moreover, for victims, the cost of being attacked continues to rise. The average cost of a data breach rose by nearly 10% year on year in 2021 to $4.24 million, the largest single annual increase in 7 years, according to IBM's Annual Cost of Data Breach Report 2021. In addition, the longer that systems remain compromised the more the costs mount.

Across industries, corporate approaches to cyber security are maturing, with companies acquiring new capabilities and bolstering their resilience. Banking and healthcare are among the most mature industries when it comes to cyber resilience, while fashion has a long way to catch up. In response, fashion decision-makers need to adopt a dual mindset, reconciling short-term needs created by the pandemic with the longer-term demands of the digital economy. To increase resilience, security should be embedded into products and processes, while customers, partners, third parties, and regulators should also be incorporated into enterprise-resilience management.

The rewards for doing so are clear for decision-makers: there is a direct relationship between cyber resilience and business performance. According to a recent McKinsey survey, higher cyber security maturity correlates with better margins, so the payoff from strong risk management extends beyond security. A successful roll out of improved cyber capabilities should be predicated on action across five key areas:

6.1.1.1 Identify the Playing Field and Risk Environment

Cyber security leaders should focus on identifying relevant cyber risks (including potential "black swan" events) across their value chains. That starts with understanding legal and regulatory ground rules, and moving to a risk-based approach. This recognizes that not all assets are created equal, and not all can be equally protected. It is vital for business leaders to take a global view of both the company's operations and its supply chains, and to communicate cyber security requirements [4] to suppliers and third parties. Insurance against cyber attacks is an option, but it is worth reading the small print; there are likely to be areas of risk that are not covered, and market conditions are changing rapidly.

6.1.1.2 Build Capabilities to Prevent Cyber Attacks

Rules and standards should be developed (such as acceptable use policies for email and anti-phishing guidelines) and technical prevention measures should be deployed across systems, including data encryption and next-generation firewalls. While some systems may need an extra level of protection, a general baseline is essential, such as keeping software up to date and regularly scanning systems for vulnerabilities. Where the cyber risk extends to production and manufacturing systems or other connected devices, measures should be expanded into those areas, too.

6.1.1.3 Reinforce the Ability to Detect and Respond to Cyber Attacks

The traditional focus of cyber security has been on prevention, but the spotlight is now moving toward detection and response, acknowledging that attackers will inevitably succeed in breaching systems. Internally, that means closely monitoring systems and applications, as well as encouraging employees to report suspicious activities. Customers, partners, and third parties should be fully incorporated into both detection and response measures. Externally, businesses should keep a close eye on cyber threat intelligence and be on constant alert, even if their own mechanisms have not yet triggered an alarm.

6.1.1.4 Clarify Responsibilities Across the Business

Clear roles and responsibilities are vital to cyber resilience. Companies need to define what "good" looks like, who owns which part of cyber security and how relevant capabilities and skills should be developed. It is essential for the company's front-line personnel and anyone who is not an IT or security professional to understand their role in identifying and mitigating cyber risk [5], and to know what level of support they can rely on. Some companies have created the role of chief information security officer (CISO), an executive who defines and leads the overarching approach to cyber security, establishes central cyber security capabilities, and helps to build capabilities across the business. While companies will need to build in-house capabilities in certain areas, they can also consider external support.

6.1.1.5 Simulate the Worst Case and Build Muscle Memory

Leading organizations test their plans and prepare for the worst by carrying out attack simulations. The aim is to assess decision-making, ensure clarity of roles and responsibilities, including decision-making power, and identify weaknesses. This enables companies to develop an effective response mechanism and improve upon their reaction speed in the event of a real attack. Companies that lead in cyber security are defined by their outstanding performance in several key areas, including maintaining a low "click rate" in employee phishing programs regularly revisiting and updating cyber security priorities; deploying solutions for managing applications; scanning the IT environment for vulnerabilities; and sourcing intelligence on threats. As an overarching principle, senior managers should incorporate cyber risk into all decision-making. In this way, they will get on the front foot and ensure the organizations' defenses are as resilient as possible.

6.2 Literature Review

6.2.1 Data Analysis

All the interviews were recorded on tape and transcribed in order to eliminate all possible errors. First, all the transcripts were read repeatedly. Then a reflection was carried out on the literature review, followed by generation of codes which allowed breakdown of data in various parts [6]. Charmaz's advice as cited by [7, 8] was taken up which cautions about ensuring the precision, simplicity, and closeness of data. The coding process of data was done in Urdu. Codes were categorized in different themes on Microsoft Excel spreadsheet, and only Urdu quotations extracted for data analysis were translated into English. Different themes were assigned to these codes in an Excel file. The coding was done through manual method. Hereafter, the codes and themes were reevaluated after keeping in view the considerations from literature review and data sample.

6.2.2 Selected Advertisement

The advertisements selected from the fashion industry were done for various reasons. Firstly, to narrow down the research frame as there is a vast variety of advertisements featuring men selling different products. Secondly, it was not possible to collect, arrange, and analyze from such a huge amount of data in a limited time, and making comparison of displays of different product types would be irrelevant. Thus, only one product type ensured that the research was specific, accurate, and meaningful.

Fashion as a topic is more accessible than any other topic among men and since everybody needs clothes to wear thus there is a likely chance that every consumer of clothes must have come across and been influenced by advertisements and fashion at least at some point in their lives. The images used by fashion advertisements selling accessories and apparel, not just serve purpose for visualization of products but also entice the viewers to imagine a fantasy flight when using that product [9]. Thirdly, fashion advertisements have a great connection with print industry and can be found on newspapers, magazines, and billboards. Therefore, fashion advertisements provided the best fit for this study due to being readily available and mostly being the source of controversy for their portrayal of gender-specific images.

In a bid to find the most appropriate print advertisements, fashion magazines from 2017 January till March 2018 were randomly shortlisted, and later scrutinized for projecting clothing advertisements targeted toward men. The print advertisements were selected from (ONE, Levis, Tony & Guy, Jockey, Dockers Khakis, Dockers). The aim was to show a variety of male depictions in advertisements to respondents in order to gain their perception. Thus, six ads were selected on the basis of classification model by [10, 11]. The classification is categorized into two variables: masculinity types and masculinity roles. Seven masculinity types were identified that were placed in four categories: macho and vigorous, sophisticated and refined, cool and trendy, and others. While masculine roles identified included entertaining, professional, decorative, recreational, and others.

While Yuan and Shaw's model has identified seven types of masculinity types, only six are used in this study due to the non-availability of the "gentle and refined" type in the three magazines that were reviewed.

6.3 Research Objective

This study targets to investigate the attitudes and perceptions of men toward depictions in printed ads in the fashion category. The objective is to explore whether male consumer segment is aware of the stereotypes present in advertisements, what they term as conventional and non-conventional, and their feelings about the tradition of gender-detailed imagery and roles.

Research question:

The research question developed to achieve the aforementioned research objective is as follows: "How do men identify gender stereotypes in printed fashion advertisements?"

6.3.1 Sub-humanization of Women in Advertisements

The products presented in various advertisements serve as the "ultimate support system" for the women in their quest to become the kind of women society approves of, and they so desire. This representation also creates a false impression that it's an innate and natural desire in women to be the best cooks and cleaners; to take excellent care of the house and its members as she does all this with an enduring grin. The women in these ads never get tired, they never get irritated, they never complain about the injustice of putting them through all this, forcing all household work on them, or for not being supported by their husbands or sons. These ads create a myth. A myth of a super-human woman. An ill-fated myth that ends up "sub-humanizing" these women in their real lives. These adverts reaffirm that the household chores are invariably associated with women, that they should look fresh and beautiful all the time, and are often laced with unrealistic body images.

A few years back, Ariel washing powder started a campaign, "share the load." The campaign was lauded by many for breaking the stereotype. The number of "progressive" ads is so abysmal in India that it catches the fancy of all the people who vouch for gender-sensitive advertising whenever it comes. Well, the Ariel ad, just like a few of its predecessors, could not make other brands to follow suit.

One cannot help but wonder why the advertisement sector is so skewed toward these old, regressive, unequal, unjust, and inappropriate societal norms. The answer lies in the very "need" of advertising the products. These adverts were made with the sole purpose of striking a chord with the audience. To leave a memory of the product in people's minds. People identify themselves with gender, thereby making communication easier. To communicate better with the audience, they focus on gender relations as practiced in society. The stereotypes and prejudices disseminated by these adverts have disastrous effects on the self-esteem of women and on their place in society. Women are forced to constantly keep judging whether they are playing the role of mothers, sisters, wives, and daughters as good as projected by these adverts. Are they fair? Are they beautiful? Are they slim enough? These trepidations exhaust women and destroy their self-confidence. This can have devastating effects on women's mental health. It may also bring self-censorship in women, limit their potential, and cost them many opportunities that may come their way. This also makes us think about the possible socioeconomic price of GA.

When the perfect reel-woman, people see on the screen, is not being mimicked by the real woman, they face ridicule, which sometimes escalates to violence against them by their partners and families. Nonetheless, what makes the whole scenario more disturbing is the fact that we are so used to these gender-based plays in the adverts that we do not feel any discomfort watching them. Usually, viewers or consumers are unaware of the toxicity, gender inequality, and regression that these adverts bring in society.

6.3.2 Advertising Standards Authority

Advertising code in the United Kingdom has banned Gendered Advertising. The Nordic Council of Ministers, presided by Finland, recently created regulation to address gender-discriminatory advertising in the Nordic countries. Human dignity is taken into consideration in many European countries like Germany and France to assess the contents of an advert. United States of America doesn't have a specific law for regulating GA in the USA. Still, other legal frameworks, like US Non-discrimination Law, are extended to cover the discriminatory advertising. In India, efforts are being made to make regulations that cater to the need to eliminate stereotypical and sexist advertising in the Indian media landscape. Gender inequality is a social issue. In addition to regulating the media space, we should also work toward changing public perceptions about gender roles to form a more equitable society.

6.4 Conclusion

The discursive strategies identified generate insights into how client organizations seek to justify and/or underplay the use of violent representations. By subverting interpretations, making authority claims and denying responsibility, the client organizations offered a range of flexible explanations in the media statements analyzed to diminish the taboo transgressions. This was enacted through two persuasive tactics that reinforced the dominance of the client organizations. First, the focus was placed upon those offended by the representations of their "incorrect" interpretations cast as uncultured, ignorant, irrational, and histrionic. Second, the role of the clients and advertisers in the events represented in the texts was minimized. Through both of these efforts, the marketers and advertisers reinforced their power as cultural intermediaries in the field of visual representation whose seeming credibility reinforced the legitimacy of their statements and actions. However, this was enacted in subtle and opaque ways that legitimatized the control of the organization but simultaneously worked to obscure the power relations at play. These discursive strategies not only fostered distance between the organizations and the transgressions but also produced anomalies that supported the crossing of such boundaries. In turn, the taboo transgressions are defended and the interests of the organizations are maintained. The following discussion considers how the state of play regarding the transgression of the taboo of violence against women through the subject positions and discursive strategies identified affects the existing social order and introduces pollution, namely the creation of uncertainties about existing gender codes in society.

References

1. Das, M. & Sharma, S. (2016). Portrayal of women in Indian TV advertisements: A study of audience perception. *The IUP Journal of Marketing Management*. Available at: http://eprints.bits-pilani.ac.in/1009/
2. Van Laer, T., de Ruyter, K., Visconti, L., & Wet-zels, M. (2014). The extended transpor-tation-imagery model: A meta-analysis of the antecedents and consequences of consumers' narrative transportation. *Journal of Consumer Research, 40*(5), 797–817. https://doi.org/10.1086/673383
3. Kim, J., Lloyd, S., & Cervellon, M. (2016). Nar-rative-transportation storylines in luxury brand advertising: Motivating consumer engagement. *Journal of Business Research, 69*(1), 304–313. https://doi.org/10.1016/j.jbus-res.2015.08.02
4. Fionda, A., & Moore, C. (2009). The anatomy of the luxury fashion brand. *Journal of Brand Management, 16*(5–6), 347–363. https://doi.org/10.1057/bm.2008.45
5. Freire, N. (2014). When luxury advertising adds the identitary values of luxury: A semiotic analysis. *Journal of Business Research, 67*(12), 2666–2675. https://doi.org/10.1016/j.jbus-res.2014.04.004
6. Bryman, A., & Bell, E. (2011). *Business research methods*. Oxford University Press.
7. Easterby-Smith, M. T. (2012). *Management research*. London: Sage.
8. Hansen, J., & Wanke, M. (2011). The abstract-ness of luxury. *Journal of Economic Psychology, 32*(5), 789–796. https://doi.org/10.1016/j.joep.2011.05.005

9. Santaella, M., Summers, T. A., & Kuttruff, J. T. (2014). Involvement in fashion advertising: Image versus text. *Academy of Business Journal, 1*, 8–23.
10. Shaw, S. M. (1994). Gender, leisure, and constraint: Towards a framework for the analysis of women's leisure. *Journal of Leisure Research, 26*(1), 8–22.
11. Al-Ibrahim, A. (2014). Quality management and its role in improving service quality in public sector. *Journal of Business and Management Sciences, 2*(6), 123–147. https://doi.org/10.12691/jbms-2-6-1

Chapter 7
Trolls to Cyber Mob: Reasons of Trolling on Women

C. Karthika

7.1 Introduction

The internet enables communication devices to become part of everyday life, life, especially for children and young people. They use it as part of their academic work and to get connected to friends all-around through social media [1]. The technical functionality of screen-based devices has become more interactive and persuasive since the introduction of iPhone in 2007 and android in 2008. This increases the use of smartphones among youth [2–4].

When the entire world is shrinking into the internet, real-life misogyny also spills into the virtual world. From verbal abuse to obscene videos and morphed photos, there are many ways in which men harass women in cyberspace. According to the National Women's Commission, the total number of cybercrime complaints received is 704 with a minimum number in February (21) and a maximum in July (the situation that is not different). The police registered 2019 cases under the IT Act. Surprisingly, the count was much lower than that of 2018, when there were 346 cases registered. Many say complaints lodged by women against cybercrimes evoke little or no response at all from the police in Kerala [5]. The research by Broadband Commission Working Group on Gender found that 73 percent of women across the globe had already faced online violence then [6]. The study also inferred that women aged 18 to 24 were at a heightened risk of being exposed to every kind of cyber violence ranging from stalking and sexual harassment to physical threats.

C. Karthika (✉)
Department of Visual Media and Communication, Amrita Vishwa Vidyapeetham, Kochi, India

D. Mishra et al. (eds.), *Communication Technology and Gender Violence*, Signals and Communication Technology, https://doi.org/10.1007/978-3-031-45237-6_7

7.1.1 Cyber Violence

Cyber violence can be considered as the evident index of the spectrum of sexism and sexual harassment that exists within our society, part of Kelly's "continuum of violence" [7]. The advancements and inventions in the field of communication technologies have bought new tools and venues of violence against women, especially young women. The atrocities against women in any society are carried over to a new venue, i.e., cyberspace where they are subject to a range of abuse and harassment. In short, violence against women is reproduced in cyberspace.

Weinstein and Selman said that online pantomime, individual assaults, open disgracing, and mortification typify social "hardness and ruthlessness" attacking the computerized space. So, advanced apparatuses offer incredible, yet harming courses for youngsters to convey and react. Some studies bring forward the concept of "fraping", which normalizes online interactions which are violent and sexual, and belittles the act of rape [4].

Mason (p.323) defined cyber victimization as "an individual or a group willfully using information and communication involving electronic technologies to facilitate deliberate and repeated harassment or threat to another individual or group by sending or posting cruel text and/or graphics using technological means" [8]. Unlike traditional victimization, there is no face-to-face interaction which gives courage to the offenders to engage in such activities increasingly. Here, gender is an inevitable factor [9, 10].

Cyber violence shows itself in myriad forms: hate speech; stalking; hacking; image manipulation; threats; privacy violation; cyber bullying and online pornography. The internet more generally facilitates trafficking and the sex trade. Women are guilty of many of these forms of abuse (their targets are often other women, which is how hegemonic. Institutions such as patriarchy work), just as they are in wider society, but most of these abuses are committed by men and we should not overlook that fact.

7.1.2 Gender Trolling

According to the researchers, the term "trolling"refers to "inadvertently misleading, disruptive, and destructive behaviour in online social situations" [11, 12]. On the other hand, a specific person may be the victim of online stalking or other forms of harassing correspondence [13]. Trolling is defined as "the act of purposefully generating emotional responses from persons on the internet through provocative or offensive rhetoric". Trolling topics can range from politics and current affairs to celebrity and sports news, all with the goal of generating annoyance to their targets for several reasons [14].

While acknowledging that the relationship between black humour and online harassment is complex, the goal of this article is to investigate trolling as a broad concept that encompasses anything from "ludicrous rants to inane thread jackings to personal insults and abusive language" when people interact online. As a result of this choice, it is now possible to investigate how female journalists deal with a variety of oppositional behaviours, some of which are unpleasant and others of which are hostile.

Trolls, like the renowned beast from Scandinavian folklore, are now a staple of the internet, according to certain authorities. Trolls don't care about their social class, caste, gender, or job; they are all on an equal playing field. It depends on various circumstances whether you are subjected to trolling. Women and celebrities, for example, are more likely than men to report serious abuse, sexual innuendos, rape threats, graphic statements about their bodies, and body shaming, whereas men are less likely to disclose these incidents [15].

Rape and murder threats are the most common kinds of gender trolling directed towards women, ranking first and second, respectively. Rape culture has returned in public debate in the last 5 years, with trolls employing rape threats as a regular lexicon. It is a complex system of beliefs that condones physical and mental terrorism against women and presents it as the norm, according to the explanation of what develops male sexual aggression and supports violence against women.

Some people are now threatening a female commenter with rape to express their displeasure with the situation [16]. This phenomenon, which is responsible for both transmitting and perpetuating the existing rape culture, is also responsible for the speed with which photos and written statements are distributed on the internet [17]. Students were polled via social media about their interactions with rape culture and how they reacted to it. Sexist and misogynistic words and behaviours directed at women are promoted in the context of "lad culture", which includes sites such as the "Lad Bible" as well as racist and sexist trolling on social media platforms. Trolling is a part of everyday life in today's society and culture. Trolling can be used to silence women's voices in online and virtual public places (or in settings where women are not permitted to speak), contributing to the heteronormative masculinization of online space.

Previous research has found a trend in how women respond to abuse when it occurs outside the family. The advice of James E. Gruber on dealing with abuse has resulted in the development of four distinct solutions [18]. Avoidance, diffusion, negotiation, and confrontation, to mention a few, are all useful methods.

7.1.3 Problem Statement

This research intends to throw light on the reasons of gender-based trolls on young women. It draws a link between gender trolling, cyber violence, cyber victimization, and psychological well-being of the respondents.

7.2 Research Design and Methods

The study uses qualitative methods. The responses were collected from three focus group discussion to find the impact of gender trolling on women in Kerala. The samples were identified using purposive and convenient sampling methods. The names of the respondents will not be revealed to assure the anonymity.

There were 101 participants in total, with women accounting for more than half of them. In the first focus group there were 35 respondents, all females, in the second there were another 30 respondents, all males. In the third the focus group was mixed with both male and female respondents. The participants ranged in age from 21 to 68, with an average age of 41. While respondents came from all throughout Kerala, the cities with the biggest proportion of respondents were Kochi and Trivandrum.

7.3 Results and Discussions

7.3.1 Observations on Gender Trolling

As part of the study, participants were asked if they had or would have any concerns about being attacked on social media for being women. The study's goal was to find out how people felt about cyber bullying. In the survey, the respondents highlight reasonable concerns in the misogyny comments and even trolls and posts by the opposite gender about their body, political stands and even on a normal photograph. A bit more than half (50.7 percent) of those polled expressed fear about being targeted by internet trolls. The women were the ones who were the tensest. Women were more concerned than males about the frequency of online misogyny, according to the survey.

Because they viewed social media attacks as an expected part of the online experience, most participants expressed no concern about them. People were not deterred by internet trolls in most cases. Five respondents stated that these gendered trolls demotivated them to be more active on social media. These women are concerned about the potential harm that such attacks could cause to their reputations, mental health, and relationships with their families. This study's findings, despite the small sample size, demonstrate that social media attacks are gendered and, if efforts to combat online hostility are unsuccessful, may constitute a gender barrier. The causes for misogynistic attacks against women on social media are twofold.

7.3.2 Responses on Gender Trolling

The experiences of females subjected to gender trolling should serve as a sobering lesson. They realized that if they wanted to be active in social media, they would have to be prepared to deal with gender trolls. But not all women are willing to put up with sexist remarks, as evidenced by the following statistics. Many people believe that contemporary women leaders can contribute to the prevention of gender-based harassment and the increased accessibility of online for future generations of women. When a lady applied for a faculty role in a university, she was harassed on the internet by trolls. Her argument is that "more women need to get out there and speak up for what's right", and she goes on to say that the current system is sexist. "It can't be prevented only through mere awareness programmes, the law must be strong", a lady participant responded. "More women should come forward, open. This will make the victims courageous", said another respondent. "It is not just the case of simple trolls in social media. It has several dimensions. It can even lead to the disappearance of the victims from social me said a lady. Body shaming, caste, political stand, family, friendship … and whatnot. A woman is audited through several milestones in social media", and responses go like this.

7.3.3 Dangers of Gender Trolling

Anonymity has been blamed for the manipulation of internet discourse. A person's accuser can be determined by staring them in the eyes during a face-to-face meeting. In normal, face-to-face communication, many people are reluctant to express themselves in emotionally charged, profane, or even offensive language. This barrier is removed by the anonymity of social media. Because they use fictitious names and post under pseudonyms, anonymous commentators can say whatever they want without fear of being held accountable. "It's simpler to attack from behind the screen", a female respondent explained. Because they do not have a name or an identity, they can say whatever they want. Trolls'use of anonymity makes it more difficult for victims to reply to or stop the attacks. This simply serves to fuel the trolls'fire. According to a study, the greater the belief that internet commentators are anonymous and unaffected by the consequences of their behaviour, the more likely they are to engage in cyberbullying [19]. However, not all the participants in the research were opposed to internet anonymity. In a statement, a male respondent supported anonymously blogging online. He was worried that the right to free expression was being "slowly but steadily destroyed". People like him can have a public voice if they maintain their anonymity, even if their beliefs are contentious or even offensive. He expresses his opinions anonymously because he is terrified of losing business if people disagree with him. Women have been accused of being targeted by online trolls, which participants expressed worry over. According to Mantilla, many of the women's responses to social media attacks fit the concept of gender trolling. He feels that women are harassed online simply because they are women, not because they have done anything wrong. The respondents from both genders have been enraged by internet trolls who objectify women by focusing on their body parts and threatening them with sexual threats. "If you are overweight, how can you know?" What do you know, you're only a female are two other options. Trolls on the internet are posted by people who are sitting in their undies in a dungeon somewhere, without a life of their own. After seeing or hearing [about me], some individuals will go from "You have no idea what you're talking about" to "You are a dumb lady". According to participants, some lie at the root of online harassment and denial of feminist goals. Racism, homophobia, and/or Islamophobia were also prevalent themes in gender trolling. The goal is to disenfranchise everyone on the internet, regardless of how a woman's skin colour or sexual orientation influences the nature of an assault. Women's gender trolling comments imply an apparent link between online abuse of women and social misogyny. Gender trolling, according to a respondent is more common among women than among men. "There is a segment of society that targets women in that way, and there is an element of it in every society. He made a remark. You never know when it will be over," a female respondent said. As previously noted, "feminists and feminist viewpoints are the targets of a significant amount of online harassment".

7.3.4 Trolling as an Organized Discourse

New concerns have emerged for its users due to censorship, surveillance, and control over the content and the bodies and online avatars of those who use online platforms. While trolling can be "deceptive, damaging, or disruptive", gender trolling takes it a step further by specifically threatening and harassing women. The hazards and abuses associated with a more sexist, hierarchical, and politicized environment become more numerous and severe. Trolling technique evolves in direct proportion to the trolling objective. When trolling takes on the form of a focused, orchestrated, and purposeful attack on a particular ideology or group of beliefs, it becomes harassment. When a joke is used to silence purposefully, muffle, and obscure people who disagree with it, it transcends the level of a simple jest. Rape and gender-based violence are particularly deadly for women and anyone who identify differently, such as members of weak castes or minorities. Rape and gender-based violence are grave dangers to one's right to free expression. Additionally, serving as a deterrent, it has major effects on individuals. Trolling and disciplinary language, according to Cole, "individualize bodies that are dispersed and dispersed in a network of relations". Additionally, she believes that trolls discourage and punish women in online forums by "threaten[ing] her body with violence". An examination of the content and forms of trolling can shed light on how the body, particularly the body of a woman or other marginalized gender, can be used as a means of control, enslavement, and surveillance. According to Michel Foucault, this is the new "political technology of the body" that has resulted in the emergence of new kinds of power and exploitation. As a result, the question of whether the technology would result in increased dystopia in society has become increasingly convoluted, as it simultaneously threatens and supports the world's current status quo.

7.4 Conclusion

Women are significantly more prevalent in Indian society than they were a decade ago. The proportion of accomplished female employees who are active on social media has increased. In contrast, concerns about how to manage abusive behaviour, harassment, and biased speech on social media will become more prevalent as their use increases. As a result of women being socialized to embrace these practices as normal in the workplace, their prevalence has increased to the point where they are the norm. The internet-based culture of the twenty-first century has normalized the practice of reading and responding to objectionable comments made by the public. Although gender is not a factor in trolling, it is evident that in patriarchal nations such as India, references to female targets frequently contain sexist connotations. Typically, females respond to cyberbullying by blocking or ignoring users who make sexist comments. Alternatively, some avoid posting about topics that could promote abuse. Despite this, few individuals engage in combat or even attempt to defend themselves.

In conclusion, in the rapidly evolving digital media environment of the present day, women are socialized to expect and value the inclusion of a diversity of distinct perspectives in their daily work. They view baiting as a necessary addition to their predetermined daily work schedules. Despite the possibility that this form of bullying does not impede "mass self-communication", the results of the study indicate that it does result in self-censorship.

Trolling has evolved from harmless online banter to harassment. Trolling has been elevated to the level of an intentional effort to confuse, repress, and stifle opposing viewpoints. Fear of rape, gender-based violence against women, and other identities, such as belonging to a vulnerable caste or community, are instances in which trolling poses a genuine threat to life and the right to free speech. To intimidate and subjugate women online, it is found that the perpetrators "threaten her body with violence". As we analyse the various forms and contents of trolling, we can gain a better understanding of how the body, and specifically the body of a female or other marginalized gender, can be a contentious site of control, subjugation, and surveillance. The fact that cyberspace maintains and challenges the status quo has made it a controversial topic, further clouding whether technology will lead to social dystopia.

References

1. Ngantcha, M., Janssen, E., Godeau, E., Ehlinger, V., Le-Nezet, O., Beck, F., & Spilka, S. (2018). Revisiting factors associated with screen time media use: A structural study among school-aged adolescents. *Journal of Physical Activity and Health, 15*(6), 448–456. https://doi.org/10.1123/jpah.2017-0272
2. Ofcom, U. (2016). *Children and parents: Media use and attitudes report.* Office of Communications London. Mascheroni, G, & Cuman, A. (2014) *Net children go Mobile: Final report.*.

3. Ofcom, U. (2017). *Children and parents: Media use and attitudes report*. Office of Communications London.Daring act of activists a wakeup call for cyber police. *The New Indian Express*. October 1, 2020.
4. UN Women from Insights to Action: Gender Equality in the ... - Headquarters. Accessed 27 May 2022. https://www.unwomen.org/sites/default/files/Headquarters/Attachments/Sections/Library/Publications/2020/Gender-equality-in-the-wake-of-COVID-19-References-en.pdf
5. Kelly, L. (1988). *Surviving sexual violence*. Polity Press.
6. Weinstein, E. C., & Selman, R. L. (2014). Digital stress: Adolescents' personal accounts. *New Media & Society, 18*(3), 391–409. https://doi.org/10.1177/1461444814543989
7. Dehue, F., Bolman, C., & Völlink, T. (2008). Cyberbullying: Youngsters' experiences and parental perception. *Cyberpsychology & Behavior, 11*(2), 217–223. https://doi.org/10.1089/cpb.2007.0008
8. Lenhardt, A. (2020). *Teens, Smartphones & Texting*. Pew Research Center: Internet, Science & Tech. Pew Research Center, August 27, 2020. https://www.pewresearch.org/internet/2012/03/19/teens-smartphones-texting/
9. Troy McEwan Lecturer in Clinical and Forensic Psychology. Personality differences: Trolls and cyberstalkers aren't the same. *The Conversation*, June 6, 2022. http://theconversation.com/personality-differences-trolls-and-cyberstalkers-arent-the-same-23309
10. Buckels, E. E., Trapnell, P. D., & Paulhus, D. L. (2014). Trolls just want to have fun. *Personality and Individual Differences, 67*, 97–102. https://doi.org/10.1016/j.paid.2014.01.016
11. Burcham, F. J. (2016). Trolling: How to combat online harassment. *Business 2 Community*, December 9, 2016. http://www.business2community.com/cybersecurity/trolling-combat-online-harassment-01726624
12. Nair, S.. (2017). The Sanskari troll and three other types of online pests who love to attack Indian women. YourStory.com. *YourStory*, February 28, 2017. https://yourstory.com/2017/02/online-trolls/amp
13. Jane, E. A. (2012). Your a ugly, whorish, slut. *Feminist Media Studies, 14*(4), 531–546. https://doi.org/10.1080/14680777.2012.741073
14. Steeves, V. M., Bailey, J., Shariff, S., & DeMartini, A. (2015). "Defining the legal lines: EGirls and intimate images." Essay. In *EGirls, ECitizens*. University of Ottawa Press.
15. Sills, S., Pickens, C., Beach, K., Jones, L., Calder-Dawe, O., Benton-Greig, P., & Gavey, N. (2016). Rape culture and social media: Young critics and a feminist counterpublic. *Feminist Media Studies, 16*(6), 935–951. https://doi.org/10.1080/14680777.2015.1137962
16. Barlett, C. P., Gentile, D. A., & Chew, C. (2016). Predicting cyberbullying from anonymity. *Psychology of Popular Media Culture, 5*(2), 171–180. https://doi.org/10.1037/ppm0000055
17. Mantilla, K. (2015). *Gendertrolling: How misogyny went viral*. Praeger.
18. Cole, K. K. (2015). 'It's like She's eager to be verbally abused': Twitter, trolls, and (En)gendering disciplinary rhetoric. *Feminist Media Studies, 15*(2), 356–358. https://doi.org/10.1080/14680777.2015.1008750
19. Foucault, M., & Sheridan, A. (2020). *Discipline and punish: The birth of the prison*. Penguin Books.

Chapter 8
Gender Biasness – A Victim of Artificial Intelligence-Based Development

Sonal Pathak, Vijender Kumar Solanki, and Nguyen Thi Dieu Linh

8.1 Introduction

Today we are living in a world of high possibilities and smart work styles. Only hard work cannot pay off these days. The time has come when a human is working with machines or vice versa to increase productivity and long-term profits. The involvement of machines is not new, simple computer has already made their place a long way back. Knowingly or unknowingly Information Technology industry had become dependent on Artificial Intelligence technologies. Gradually when dependencies increased, biases entered into decision models. As per the researchers, there are three major category types of Artificial Intelligence bias – negative legacy, algorithm prejudice, and underestimation. Negative legacy means biases that are the result of tempered or errors in the data input. Algorithm prejudice means the biases due to statistical dependencies of information on protected features which has been used to make final decisions. UNESCO Report, 2019 [1] discloses that gender bias has been found in training data sets of Artificial Intelligence. Unambiguously biases have been found in algorithms and this bias has replicated itself and spread reinforced gender stereotypes. These risks of gender biases minimized the role of women on a global scale and stigmatized it. This ubiquity of society has put less percentage of women in all fields – be it political, economic, or social life. In the recent past, this has hindered the progress of gender equality at the country level.

S. Pathak
Manav Rachna International Institute of Research and Studies, Faridabad, India

V. K. Solanki (✉)
CMR Institute of Technology, Hyderabad, India

N. T. D. Linh
Hanoi Institute of Technology, Hanoi, Vietnam
e-mail: nguyen.linh@haui.edu.vn

D. Mishra et al. (eds.), *Communication Technology and Gender Violence*, Signals and Communication Technology, https://doi.org/10.1007/978-3-031-45237-6_8

Many steps have been taken to remove bias from learning algorithms but they have ignored studies on how the ideology of gender has been entrenched in the language. Biased algorithms can be amended by incorporating results of decades' studies into approaches to machine learning from the text can help in removing gender biases.

Leading thinkers in the emerging research addressing bias in Artificial Intelligence are also primarily female, suggesting that those who are potentially affected by bias are more likely to see, understand, and attempt to resolve it. Gender balance in machine learning is, therefore, crucial to prevent algorithms from perpetuating gender ideologies that disadvantage women. People were using critical theory to avoid biases in personal experiences and finalized decisions. However, observing data that is present initially should be given significant importance because machine intelligence works on this initial data only. As machine learning processes big volume data, if this data would have been laden with the stereotype framework of gender, technology would generate a biased resulting application.

8.2 Gender Biases

If a person is being treated or receiving a behavior on the person's identity based on gender. All the aspects and concerns related to men's and women's lives and the societal situation can be summarized by the gender of an individual. In a society, how a person interrelates, differences in access, use of resources, reaction to change, policies and interventions, and their activities. In society, each gender has a significant role in access to privilege, power, and possibilities that one gender has, and not given to the other gives birth to gender biases. It escalates discrimination and gender inequality. An example of gender biases can be the specification of a person by seeing their occupation of doctor or engineer as "Male". Similarly, a teacher profile can be assumed to be "Female". This stereotype mentality is common to see in any society. Generally, female has restricted admittance to the labor market. Especially in countries like India, a very less percentage of female youth enjoy the privileges given to youth males. Women have admitted that they suffer this gender discrimination from their families only by getting the lesser opportunity for education, freedom from household chores, etc. Moreover, female youth is not aware that they are underprivileged due to their gender as well.

8.3 Women and the Industry

In the Industrial sector, the percentage of female workers' participation varies not only in different industries but also in different states of India. Research studies by Fajimi and Omonona [2] reveal that because of poverty, the proportion of women workers increases in the workforce. This study reveals that a high proportion of poor people in a state can be estimated by a higher proportion of women in the workforce in that state. Gender differences can be highlighted in terms of wage differences of each gender as well.

Perinelli and Beken [3] found that female workers are getting lesser wages as compared to male workers irrespective of geographical regions and sectors. Even sectors like manufacturing and services sectors and agriculture sectors are revealing this fact about female workers.

Swift [4] report depicts that though a high number of women graduates are there in Europe, there are less number of female workers in the market which is associated with labor. Even with the same education, male workers are getting more wages than female workers which leads to a "gender wage gap".

Machin and Puhani [5] found that differences in the geographical regions cannot change the gender wage gap analysis. Countries like Pakistan lag far behind in terms of gender equality. Women are more focused on lower-level or middle-level work where mostly unskilled workers are required.

One of the other reasons for the lesser participation of women in the workforce is directly related to their dependencies on family for their day-to-day decisions, their mobility from one place to another place, and restricted access to money as shown in Fig. 8.1. Despite household practices that favored male youth over female youth, mobility, independent decision-making, and access to money were not universal among male youth. Fewer than half of male youth could express their opinion to elders (aside from parents) or confront others who had wronged them. Most male youths did not have access to money. Thus, there is considerable room for improving independent behavior among male youth as well.

Fig. 8.1 Workforce diversity by gender

Shidhaye [6] found that females adopt labor-saving technologies at a slower pace in comparison to men. They concluded that reducing female's work is the effect of technologies in both domestic and productive tasks. Affordability, physical accessibility, and cultural appropriateness are the major reasons for gender differences in industries, especially the IT industry. As informal sectors are growing faster in this economic development phase, the engagement of females is more in the informal sectors. Another sector in which more percentage of women is the service sector jobs.

Linden et al. [7] The share of women employees is very low in the emerging services of IT sectors. The share of women workers in the software industry is 27%. More percentage of women's workers lie in the lower-skilled and lower-end jobs. One of the fields where the share of female employment is more is Agriculture.

Dastin [8] reveals through a report that more women are employed in silk-sericulture and taking care of their families through this employment. They are helping their husband in their occupation and educating their kids but in IT sectors, this percentage is very low.

Vincent [9] explored that 72.70% of women was the part of employment which was generated by the establishment of a mulberry garden to cultivate and rearing of silkworms for the production of cocoons to produce raw silk. More women labor can be observed in egg incubation, sewing, and, cocoon cutting but a very less percentage of women is observed in the IT-industry, electronics, cyber security and such sectors. One more factor that exists in the IT industry is that women do not opt for night shifts. Special arrangements or measures have to be provided to enable safe working environment for women in the night shifts.

8.3.1 Women in the Field of Cybersecurity

If we look for the industries wherein the gender gap persists, no doubt "Cybersecurity" is one of such fields. Historically and currently, males are dominating the field of cybersecurity. There is no surprise to know that in cybersecurity jobs, women are underrepresented as information technology has been considered stereotypically male-dominated. To fill this gap more women-dominated positions in the cybersecurity field and finally leadership position is required. For this misconceptions about the cybersecurity field should be rectified. Those who have succeeded in this field reveal that cybersecurity is not only about technology but also it's about tracking the data and giving protection to people who have compromised their data. Those people who see thing differently and wants to do something different and look for success in this fast-paced environment should enter this field of cybersecurity – Be they Female or Male!

8.3.2 *Present Scenario*

As per a recent study, only 14% of the total cybersecurity workforce are female employees. Surprisingly only 1% of these 14% female cybersecurity workforce possess any senior position or leadership authority. This shows that the new outlook workforce in cybersecurity pertaining pay disparity between male and female employees because counting of men outnumbered the counting of women. ISC report says that though women's participation in the cybersecurity field is growing gradually still women make up a very lesser percentage of the workforce and this increment is not enough. Cybercrime Magazine (March 2021) discloses that women now make up approximately 20% of the global Infosys payrolls. This is, of course, an encouraging improvement from six years ago, but still woefully shy of the 50% range that would represent parity.

According to Cybersecurity Ventures, in comparison to 10% in 2013 and 20% in 2019, women are holding now approx. 25% of cybersecurity jobs in 2021 at the global level as shown in Fig. 8.2. The research is considering corporate network jobs and includes IIOT and ICS security, OT security, aviation cybersecurity, automotive cybersecurity, medical device security, and other market categories. Though now a day, women are making their place in the cybersecurity service provider ecosystem very well which also include digital forensics and small business owned by women. A prediction made by cybersecurity ventures reveals that women are going to represent 30% of the global cybersecurity workforce by 2025 and in 2031 it will reach 35%.

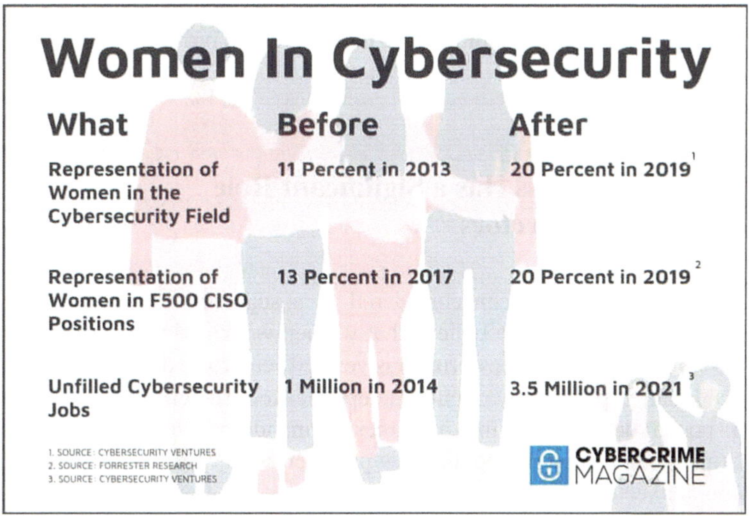

Fig. 8.2 Women in cybersecurity. (Source: Cybercrime Magazine, 2021)

8.3.2.1 Reason for Lesser Number of Women Representation

Gender Stereotype

The notion that women cannot make their career in the field of cybersecurity may be false but somewhere it influences society and our young girls. They are told that it is very difficult to make and sustain your career as a techie in network security. If, however, girls choose their career in the IT security team by breaking these barriers may find themselves surrounded by men and an unconscious bias makes a woman find opportunities for career advancement. This stereotypical idea that cybersecurity is solely a technical field and men are better suited for a network security job is influencing the entry of women workforce in cybersecurity. because of such baseless notions, women are steered away from the opportunity, classes, and training that can lead to careers in the cybersecurity field. There are a few myths that exist in society about women making a career in the cybersecurity field.

The First Myth-

American Psychological Association reported that men and women have equal capabilities in dealing with math and science subjects and even in verbal ability. It is a myth that men are more competent in dealing with practical subjects or dealing with technology and network security issues and in the consequences boys are given more opportunities to get exposure to practical ability-based job options and vice versa than girls. Thus if women assume that the cybersecurity field is not for them, it is based on purely stereotypical mindsets and not based on reality.

The Second Myth-

The notion that only strong technical skills are required in the cybersecurity field needs to be examined once again closely as cybersecurity is a field where a variety of perspectives is required and innovative methods are to protect the data in terms of creating passwords, etc.

8.4 Gender Biasness Has a Significant Role in Corporate Sectors

The corporate sector has been considered as a segment where more liberty is expected for each gender. We believe that women working in the corporate sector are more empowered but empowerment is very difficult to be described in a nutshell as it is a dynamic process. Economic independence, self-reliance, social transformation, power decision-making resources, demanding equality, and knowledge enhancement can be considered as a few parameters of women's empowerment in the corporate sector. Thus the need for hours is the generation of a socio-cultural environment that can help in minimizing gender-biased activities and enhance rational behavior which can balance the role of men and women in the corporate sector. Thus to empower women the corporate sector gender sensitization must be taken as an essential policy – be it at the senior level or middle level so that policymakers can also take care of gender balance practices while framing new policies on which training data for an algorithm of decision-making model can be dependent. The contribution of men will only be recognized when men change their perception. To remove gender biases from training data, gender sensitization at all vessels of the organization is mandatory. Women's talents can be nurtured and control over conscious efforts can only be seen if gender sensitization is promoted. Legal, health empowerment of women, social, political, and educational growth of women are only possible in the corporate sector if gender sensitization is promoted.

It can be concluded that women comprise around half of the human population, but Indian society is still male-dominated and women are not considered equal partners, either inside or outside the four walls of the house. They are treated as weak and dependent creatures. Although the government is taking various initiatives to promote women's empowerment, the actual results were not observed. The required quantum of empowerment is still a dream. The need is to reach empowerment by a gender sensitization approach. We need to involve both men and women in the gender sensitization process so that men are mentally prepared for delicate roles, responsibilities, and power over their female counterparts. Lack of gender sensitization was the reason for not achieving much despite many efforts since independence. The need of the hour is to change insight into the whole society so that it recognizes women with a positive approach. Only when the men accommodate women by shading their ego, actions and efforts will be visible. Gender sensitization refers to the sensitization about gender equality concerns. It helps people in examining their attitudes and beliefs and questioning the realities of both sexes. Issues such as the problem of sexual harassment, sexual stereotyping, dearth of counselor sensitivity to complexities such as caste, minority experience or sexuality, and lack of special emotional and academic support both at the personal and institutional level for students from marginalized groups.

8.5 Machine Learning – A Type of Artificial Intelligence

Machine learning (ML) is a type of Artificial Intelligence that allows software applications to become more accurate at predicting outcomes without being explicitly programmed to do so. Machine learning algorithms use historical data as input to predict new output values.

Recommendation engines are a common use case for machine learning. Other popular uses include fraud detection, spam filtering, malware threat detection, business process automation (BPA), and predictive maintenance. Machine learning is important because it gives enterprises a view of trends in customer behavior and business operational patterns, as well as supports the development of new products. Many of today's leading companies, such as Facebook, Google, and Uber, make machine learning a central part of their operations. Machine learning has become a significant competitive differentiator for many companies.

Process of Machine Learning/Artificial Intelligence
Classical machine learning is often categorized by how an algorithm learns to become more accurate in its predictions as shown in Fig. 8.3. There are four basic approaches: supervised learning, unsupervised learning, semi-supervised learning, and reinforcement learning. The type of algorithm data scientists choose to use depends on what type of data they want to predict.

8.6 Artificial Intelligence-Based Development Affected Gender Equality

There are several forms of Artificial Intelligence bias. One of the biases is cognitive biases which enter into models of machine learning and data set by human developers. From such entries, business gets into the algorithm. Another reason for biases is the incomplete data and this becomes significant if due to cognitive biases, this information is being omitted. By using technology, the goal of inculcating social

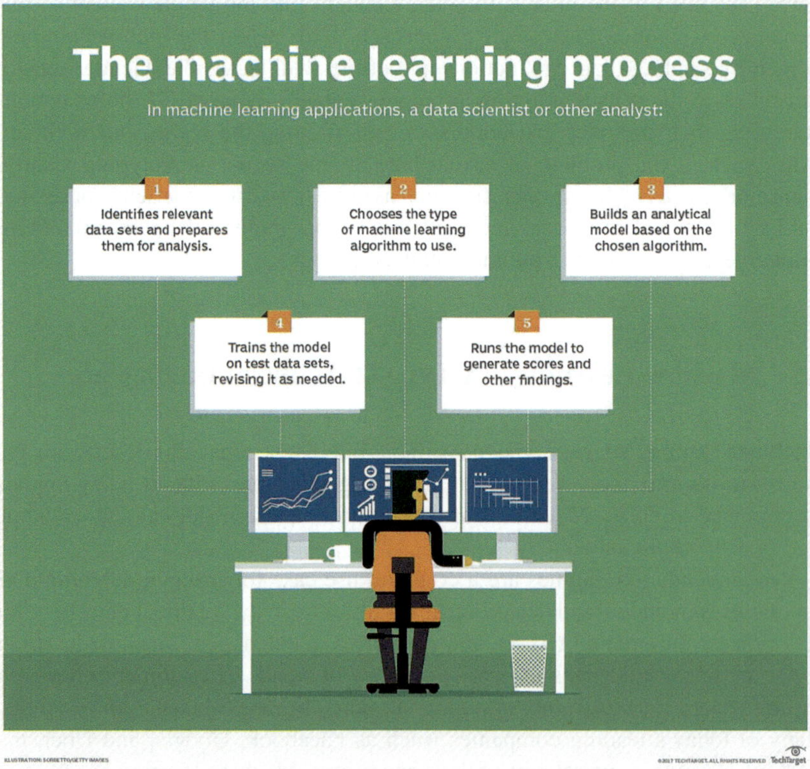

Fig. 8.3 The machine learning process

gender balance can be achieved with the amalgamation of gender balance in Artificial Intelligence applications. As now a day, Artificial Intelligence is becoming widespread, diversity in gender in data and governance of Artificial Intelligence should be utilized to promote gender inequalities and help communities that are struggling for controlling violence against gender and can inculcate a good work environment in the IT industry. These crucial issues need to be taken care of at the war front level now.

For pattern recognition, Artificial Intelligence increases applications of the algorithm, and the use of machine learning gets lots of funding and attention as well. To hold a high place as a global leader in Artificial Intelligence, the Government of Canada in 2017 announced $ 125 million in developing Artificial Intelligence strategies for Pan Canadian in the 2018 budget. The government reaffirmed its commitment of government in investing more in Artificial Intelligence-based projects. The Canadian government has made explicit strategies for promoting women empowerment and gender equality a central theme and goal of its Pan-Canadian strategy. National Artificial Intelligence Research and Development Strategies plan mentions in its policy that racial, age, economic, and gender will be the main outline for consideration of the futuristic plan of growth in the IT sector. To promote gender equality and advance women empowerment issues related to gender need to be highlighted in all sectors and disciplines including Artificial Intelligence [10]. Though Artificial Intelligence has been considered a threat to gender inequality still society cannot ignore the potential of Artificial Intelligence in bringing positive changes regarding norms of gender balance.

For Instance-One Artificial Intelligence powered recruitment system has been found with gender discrimination but Artificial Intelligence-based decision models also help employers to write job postings in which more sensitive language regarding women can be used to promote gender diversity in the workforce. Therefore, Artificial Intelligence can become part of enhancing management practices in our society to support gender equality and can minimize gender discrimination [11].

8.7 Reinforcing Gender Biasness and Inequalities Through Artificial Intelligence

The robustness of Artificial Intelligence depends on the kind of input we give to it. An algorithm reads input data to recognize a data pattern. Development of data set in terms of relevant and irrelevant data is done by data researchers on the basis of their conscious and experience. Thus the output of Artificial Intelligence is completely dependent upon the quantity and quality of inputs which is dependent on human decisions and the kind of existing data in the real world as depicted in Fig. 8.4. Thus this relationship between input and output of the algorithm is completely justifiable. It's about giving a data pattern to the algorithm.

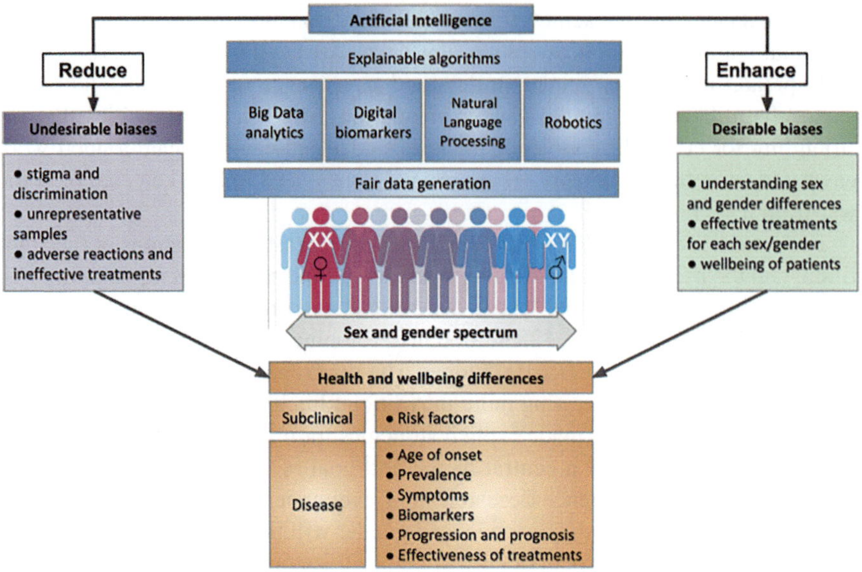

Fig. 8.4 Factors responsible to instigate gender biasness through artificial intelligence

Therefore, input data for Artificial Intelligence technologies should be simplistic and analyzed in sense of balance between each gender in all aspects. To overcome the biases, the women percentage must be increased in Technology, Science, Maths, and Engineering. Such fields are generally solely led by men. In such a scenario, input data ignores the diversity which can be there in the perception of different genders, and their considerations for some issues. As a result, gender inequality and biases get to enter into a pattern of data [12].

A big change in the culture is required to implement systematic reforms. It is unclear how Artificial Intelligence will fix this inequality of gender itself. An algorithm is only replicating its data patterns. If a company wants to figure out their customers who are premium based on their purchase and the company wants to provide some extra benefits to these customers. An Artificial Intelligence model can be utilized to figure out such customers by using details of customers as input data sets. But the quality of this input data should be ensured by removing blind spots and biases in it.

There is an important example of racial biases by the Artificial Intelligence model of GOOGLE in 2015, their photo-categorization software started creating decisions by recognizing people based on their dark/black color as "Gorilla". Later Google corrected this by using the Band-Artificial Intelligence solution by which the word "Gorilla" was removed [13].

UNESCO Report (2019) [1] reveals that gender biases exist in datasets of Artificial Intelligence and particular biases exist in training data sets. Removal of these built-in biases is essential and representation of the complete population in input data is mandatory. In particular, collection, labeling, and thus generation of data is done by humans only that goes into the Artificial Intelligence system model and sets the pattern and rules for an algorithm to make concrete predictions. Any Artificial Intelligence system gets embedded by biases at these two stages only.

Inequality in Practice

Gender disparity issues are more prominent in the fields where gender inequality is predominant such as – Military and Security. In the security field, work on Artificial Intelligence and research has more focused on applications of the military in comparison to peacebuilding.

Generally, Artificial Intelligence has two types of bias, as depicted in Fig. 8.5. One is "Data Bias" or "Algorithmic Artificial Intelligence Bias" – which is the result of the training of an algorithm by using biased data.

The next type of bias in Artificial Intelligence is the societal Artificial Intelligence bias. This bias occurs due to societal norms and assumptions which force us to create blind spots about certain expectations in our cognitive decisions.

Artificial Intelligence systems contain biases due to two reasons:

1. Cognitive biases: These are unconscious errors in thinking that affect individuals' judgments and decisions.
2. Lack of complete data: If data is not complete, it may not be representative and therefore it may include bias.

- Confirmation bias. This occurs when the person performing the data analysis wants to prove a predetermined assumption. ...
- Selection bias. This occurs when data is selected subjectively. ...
- Outliers. An outlier is an extreme data value. ...
- Overfitting and underfitting. ...
- Confounding variables.

One example of gender discriminatory outcomes biases in 2014 was when the Artificial Intelligence model was selecting applicants as the top five applicants from 100 applicants. Due to an error in trained data, it was selecting only male candidates

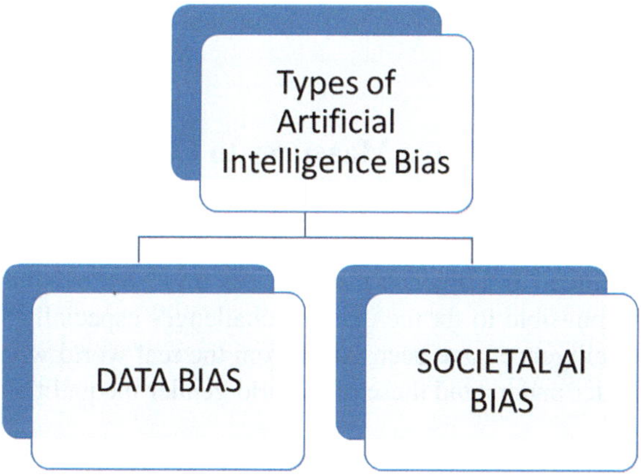

Fig. 8.5 Types of artificial intelligence biasness

selected and rejected the word "women" in the CVs. This algorithm selected mostly male applicants over a 10 10-year period. Women applicants were panelized by this biased algorithm because of the "women" word in their resumes. As soon as the bias was detected, it was removed and the company declared that they were not using this algorithm in their selection procedure.

This was a critical finding because Amazon has been using algorithmic decision-making for their recommender system to recommend customers about their purchases and products [14]. Amazon is the forefront user of such technologies as Artificial Intelligence has been a significant part of their business. To save energy and time, IT companies use these selection decision algorithms to make the process automation. Such algorithms replicated the pattern which they receive by selecting individuals. In fact, among all industries, tech companies face the highest gender disparities. 19.8% of the total workforce of IT are women programmers. Amazon learns to reinforce those normalized discriminatory outcomes. Amazon's system learns about these outcomes of normalized gender discrimination. This strategy panelized the word "women". This algorithm even downgraded the score of that candidate who has even used the word women's college in their resume [7].

The issue of gender discrimination was so deep-rooted that it could not be removed even after a few attempts to resolve it. There is no guarantee of not much happening shortly because if the system is based on a biased database, it cannot be rectified. Though it was not accepted by Amazon that it has used recommendations given by machine learning intentionally or unintentionally, it has affected the selection process.

Artificial Intelligence may affect the process of women's decision-making, therefore propositional knowledge of Artificial Intelligence must be spread. As Artificial Intelligence is increasingly used in everyday commercial applications, users may become accustomed to algorithms making decisions for them, whether they are aware that these decisions are being made or not. We may be limiting our decision-making capacities and skills when the responsibility to make decisions is relegated to Artificial Intelligence. And if Artificial Intelligence is unable to capture embodied knowledge, there are sure to be gender implications that developers and regulators alike have not yet accounted for.

8.8 Preventive Measures to Eliminate/Minimize Biasness in Artificial Intelligence

Artificial Intelligence cannot be blamed for existing gender inequalities and be responsible to fix the existing challenges especially when input data for Artificial Intelligence have been taken from the real world where such inequalities exist. To better understand these real-world gender inequalities, social innovation, and more improvised research programs should be accompanied by increased investment in Artificial Intelligence. A comprehensive analysis of the relationship between gender equality and the complexities of the Artificial Intelligence ecosystem should be practiced continuously.

There is a strong need to understand and implement the following aspects in the private and government sectors:

1. View society in a holistic view and consider the broader goals to achieve.
2. Position gender equality in Artificial Intelligence ethics and principles.
3. Design new approaches to implement principles of Artificial Intelligence and Gender Equality.
4. Identify and develop an action plan which should be beneficial for multi-stakeholders.

Women must be empowered to influence and reshape a system where Artificial Intelligence operates. Societal implications of Artificial Intelligence must be recognized not only in the technology field, but also in economics, law, security, politics, and cross-disciplinary research/education. Artificial Intelligence can reproduce and increase injustice, tribalism, and societal inequalities. Categories of men and women can be counted at the very starting point of data entry. If the difference is large, remedial measures can be taken to widen the net of more women. A gender-equitable screening process can be built by tracking this data over time.

8.8.1 Reinforcement Learning

Reinforcement learning means the training of an algorithm with a distinct goal and rules to be set to achieve that goal through that algorithm. Positive rewards can also be sought by programming the algorithm by data scientists. The ultimate goal can be achieved when it acts in the same direction. Robots are one example of reinforcement learning which perform a task based on the remaining instructions. Bots have been trained by using reinforcement learning only to respond and communicate to humans because its goal has been predefined CRM (Customer Relationship Management) software uses models based on machine learning to analyze messages in emails and respond to prompt sales. Thus we can see that a machine that is based on Artificial Intelligence and has been trained on input data can be channelized easily by using reinforcement learning. Therefore, this reinforcement learning can also train the data to balance gender equality, and the output of algorithms can be rechecked before relying on its decision model. These practices will help minimize gender biases. For example, in the HRIS system, a trained machine learning model (along with human intervention) can be used to filter applications and justify the selection of the best candidate for the job role.

8.8.2 Selection of Right Machine Learning Model

To solve a problem, the selection of the correct machine learning model is most significant to avoid Biased outputs. The following steps are suggested to minimize this hassle:

Step 1: With the help of experts and data scientists, alignment of potential data inputs and problems can be done to receive the required solution.
Step 2: In this step, data should be collected, and reformatting of data and labeling of data should be done. Execution of this step will be done by data wranglers along with data scientists.
Step 3: Select the appropriate algorithm(s) to be used and analyze its performance as well. Data scientists usually carry this step.
Step 4: Level of accuracy must be ensured by fine-tuning outputs continuously. A data scientist should perform this based on feedback from data experts in the field of Artificial intelligence and technology.

8.8.3 Usage of Human Interpretable Machine Learning

Explaining how a specific Machine Learning model works can be challenging when the model is complex. There are some vertical industries where data scientists have to use simple machine learning models because it's important for the business to explain how every decision was made. This is especially true in industries such as banking and insurance. Complex models can produce accurate predictions, but explaining to a layperson how the output was determined can be difficult. Major IT players are using the application of Artificial Intelligence in their spectrum of Machine Learning activities. Big vendors such as – Microsoft, IBM, Amazon, Google, and other IT leads are in the race to sign up customers for their services which are Artificial Intelligence-based platforms. These leads are using Artificial Intelligence for collection of data, processing of data, classification of data and building models, training, and deployment but human intervention should not be ignored at any stage of data processing. This will help in reducing error due to training of data set and will not produce unnecessary biases in the results.

8.8.4 The Right to Internet

As per the UNESCO report [1], In the Digital Economy, women share a very less share in advanced technology jobs where non-task-routine is in demand. A major problem in India is the result of a lack of access to technology. This report reveals that 56% of Indian men own a mobile phone in comparison to 46% of Indian women aged between 15 years and 65 years. For Indian Law, to enhance the rights of women in India, there should be the principle of the right to the internet for Indian Women. In 2020, mobile phones and internet access by everyone have been recognized as an everyday necessity by Kerala High Court. This should be considered a significant part of freedom of expression and speech. The United Nations Human Rights Council and the General Assembly have adopted a resolution that depicts that to enhance the quality of education and information, the right to the internet should be considered an essential aspect. Article 21 of the constitution has given two rights – Rights to Personal Liberty and the Right to Life. The government looked at the view that the right to be able to access the internet can be part of this article.

8.8.5 Role of Labor

The Institute for Women's Policy Research found that due to job automation, women are at a high risk of displacement in comparison to men. A category of jobs which is called midlevel jobs such as routine jobs and cognitive jobs are dominated by women which are a major threat to be displaced due to automation. As per the report of the world economic forum [57] 57% of such jobs are at risk of being displaced by 2026. For example, in a survey on gender balance workforce, reports say that in the gaming industry of the UK, 33% of women faced harassment due to their gender and 45% of women admit that their gender was a factor that has limited their progression in this career. This brings an insight that there is an urgent need to have improvised practices during the hiring process to increase the number of female programmers in the IT field.

As a first step toward this objective, California passed the California Senate Bill No. 826, which mandates that a minimum number of women be included on corporate boards. Today in the United States, women make up almost half (47%) of the workforce, but they hold less than one-third (28%) of the leadership positions in tech companies.

There is a need to involve gender experts and more participation of women at the foundation level to formulate the principles. Emphasis should be given to an increased number of women in board rooms and representation of women in technical roles in Tech companies at the Indian and Global Levels. A robust gender-inclusive Artificial Intelligence-based guidelines, principles, and codes of ethics must be incorporated to enable gender equality. Every company makes some policy related to fairness, transparency, and work ethics practices but none has framed any policy related to algorithm fairness and till now there is no clear definition that has explained "Fairness of Algorithm". More feminist principles like- languages, access to the internet, privacy, and information to make informed decisions should be common now in tech companies.

8.8.6 Data Sets – The Starting Point

Even if it is "Apparently" women-centric data, many data sets which are Artificial Intelligence based may lead to a biased result. A computer that retrieves information from a data set treats collected data as a single unit. This means that separate pieces of data are used to train an algorithm to predict a pattern inside the whole data set.

To create an Artificial Intelligence model, data sets are the very first step and therefore it is difficult to ensure that model is without bias. As we know that women are a heterogeneous group that faces varied realities. This includes indigenous women, women from remote and rural areas, women from a religious minority, women living with disabilities, etc. Thus they may not be always part of our initial data set. For example, the word "chamar" is not to be used on Twitter and it is punishable under the scheduled castes and tribes Act, 1989. Thus any specific word cannot be eliminated from the training of the data set but a systematic input with human intervention can resolve this problem. To avoid biases and exclusions into data sets, the relationship or intersectionality between gender and discrimination must be considered a significant aspect. Any information which is Artificial Intelligence generated always depends on predictions, patterns, and recommendations which are various parameters of accuracy, reliability, and universality of the used data. This information also gets affected by the biases of developers of algorithms and inherent assumptions of the data. If these coders or developers and designers are biased toward gender notions, then neutral results by machines cannot be expected and hidden discrimination will come into the system. The study reveals that in the field of Data Science and Artificial Intelligence, only 22% of professionals are women. Gender bias does creep into systems that are Artificial Intelligence based as women hold middle-level jobs only with less status.

8.8.7 Watch Carefully Where Can Gender Bias Creep in

IT professionals need to watch carefully at which step of Artificial Intelligence developed models, gender biases can creep into the system. For example, in NLP (Natural Language Processing) "HERS" is not read as a pronoun by widely used technologies. Once an incident happens with the Apple application in 2019 when it was found from a decision model that it was offering smaller lines to women with the same credit scores in comparison to men. The company then stated that the algorithm which was giving results was gender-blind but explained like that algorithm was intended to set the limit and inherently and unintentionally biased against women. A much better corporate governance in software applications is required for the prevention of such gender biases. This also includes the incorporation of gender equality practices and principles in the recruitment process and diversity in work culture and retention practices.

8.9 Conclusion

A computer's retrieved information is only good as the human mind behind it. This is the fundamental characteristic for unbiased results or gender equality, which should be kept in mind before implementing Artificial Intelligence solutions and training datasets. India failed in addressing the women's issues related to digital services and access to the internet needs women during the pandemic. India should look for such digital services, where women can depict more fulfilling roles. At present, we hardly find women's roles in the IT sectors such as – digital deliverable services and cloud computing. For a better digital future for all, societies should be more open to the requirement of gender minorities and co-ownership should be enabled so that undercooked communities should take a front seat at the table and can boost gender equality. In a society of free citizens, the core principle of developing any model must be based on data ethics. Different people have a different understanding of the meaning of fairness and biasness. Though some basic values or principles lack uniformity in recourse sharing and distribution to each gender equality means that everybody has given equal attention, and level of recourses to receive the same outcome. An organization never distributes the different amount of equity on the individual or group's need-based to achieve the goal. Equality and equity to minimize bias can lead to confusing results. The unfairness of one group should not be transferred to another group. For instance, a long-awaited list of cancer patients in a hospital should not be sorted based on an Artificial Intelligence model. Artificial Intelligence technologies have evolved with new opportunities and challenges. This is the time of action, though algorithms of machine learning have been there for past decades now it has reached new popularity as Artificial Intelligence has fully-fledged grown in prominence, which has brought gender biases as a dependent decision-making tool. Business decision makers should not ignore the significance of human intervention in any decision-making process and data analytics and data scientists should take precautionary measures while training data which can become the reason for biased decisions at a later stage.

References

1. UNESCO Report. (2019). Data retrieved from: https://en.unesco.org/ARTIFICIAL INTELLIGENCE-and-GE-2020
2. Fajimi, F. O., & Omonona, B. T. (2011). Women participation in agro-allied small and medium scale enterprise and its impact on poverty alleviation in Oyo State Nigeria. *International Journal of Agricultural Environment, 1*, 27–37.
3. Perinelli, B., & Beken, V. A. (2011). *The gender gap: A comparative analysis of wages in times of recession*. Wage Indicator Foundation. Retrieved from https://wageindicator.org/Wageindicatorfoundation
4. Swift, S. (2015). Gender disparities in the tech industry: The effects of gender and stereotypic ability on perceived environmental fit. In *2015 NCUR*.
5. Machin, S., & Puhani, P. (2003). *Economics Letters, 79*(3), 393–400.

6. Shidhaye, R., & Patel, V. (2010). Association of socio-economic, gender and health factors with common mental disorders in women: A population-based study of 5703 married rural women in India. *International Journal of Epidemiology, 39*(6), 1510–1521.
7. Linden, G., Smith, B., & York, J. (2003). Amazon.com recommendations: Item-to-item collaborative filtering. *IEEE Internet Computing, 7*(1), 76–80.
8. Dastin, J. (2018). Amazon scraps secret ARTIFICIAL INTELLIGENCE recruiting tool that showed bias artificial Intelligence women. Reuters. https://www.reuters.com/article/usamazon-com-jobs-automation-insight/amazon-scraps-secret-Artificial Intelligence recruiting-tool-that-showed-bias-artificial Intelligencenst-women-idUSKCN1MK 08G (2018). Accessed 24 Apr 2021.
9. Vincent, J. (2018). Amazon reportedly scraps internal ARTIFICIAL INTELLIGENCE recruiting tool that was biased artificial Intelligence women. *The Verge*. https://www.theverge.com/2018/10/10/17958784/Artificial-Intelligence-recruiting-tool-bias-amazon-report (2018). Accessed 28 Mar 2021.
10. Ministry of Panchayati Raj. (2012). *Annual Report-27*. Ministry of Panchayati Raj, Government of India.
11. Font, J. E., & Costa-Jussa, M. R. (2019). Equalizing gender biases in neural machine translation with word embedding techniques. Xiv preprint arXiv*:*1901*.03116.*
12. Munjal, R., & Kaur, J. (2019). Gender sensitization for women empowerment: A review. *Indian Journal of Economics and Development, 15*(1), 132. https://doi.org/10.5958/2322-0430.2019.00015.5
13. Leavy, S. (2018). Gender bias in artificial intelligence: The need for diversity and gender theory in machine learning. *2018 IEEE/ACM 1st international workshop on gender equality in software engineering (GE)*, 14–16.
14. Shields, M. (2015). *Women's participation in Seattle's high-tech economy*. https://smartech.gatech.edu/bitstream/handle/1853/53790/madelyn_shields_women's_participation_in_seattles_high tech_economy.pdf. Accessed 15 Aug 2021.

Chapter 9
A Comparative Analysis of Pornography Detection Models to Prevent Gender Violence

Deepanjali Mishra and Smriti Panda

9.1 Introduction

The word pornography is a conglomerate of two Ancient Greek words: πόρνη (pórnē "prostitute" originally "purchased", related to pernanai "to sell" from the Proto-Indo-European root per – "to hand over" – alluding to the notion of a person sold) and γράφειν (gráphein "a writing, recording, or description"), thus meaning depiction of prostitutes or prostitution (πορνεία porneía) [1]. The word pornography, often used as porn which is the shortened form, is said to contain explicit sexual content in the form of picture, video, or text, which is primarily intended to generate sexual arousal in the minds of the viewers. It is created purely for commercial purpose after taking the consent of the people who are performing during the presentation of the content which is meant for the consumption of adult viewing. Research has suggested that there are four broad motivations for people to use pornography, namely: "using pornography for fantasy, habitual use, mood management, and as part of a relationship" [2]. People in general view pornography for diverse reasons; ranging from a need to enrich their sexual arousal, as an aid for masturbation, to facilitate orgasm, learn about sexual techniques, reduce stress, alleviate boredom, enjoy themselves, see representation of people like themselves, explore their sexuality, know their sexual orientation improve their romantic

The original version of the chapter has been revised. A correction to this chapter can be found at https://doi.org/10.1007/978-3-031-45237-6-9_15

D. Mishra (✉)
School of Humanities, KIIT University, Bhubaneswar, Odisha, India

S. Panda
School of Computer Science and Engineering, VIT Bhopal, Bhopal, Madhya Pradesh, India

D. Mishra et al. (eds.), *Communication Technology and Gender Violence*, Signals and Communication Technology, https://doi.org/10.1007/978-3-031-45237-6_9

relationships, or simply because their partner wants them to. Studies have found that sexual function is better in women who consume pornography frequently than in women who do not. No such association has been found in men. As for pornography use to have any implications on public health, scholars have noted that pornography use does not meet the definition of a public health crisis [3]. Comparative studies have noted that "pornography consumption" and "pornography tolerance" in people are associated with their greater support for gender equality. People who support regulated pornography are found to be more egalitarian than people who support a pornography ban [4].

9.2 Impact of Pornography on Teenagers

Teenage is that age where the individual has no idea what is right or what is wrong. He or she may get into those areas which might lead to destruction or image building [5]. As per the suggestions of Cyber Experts, getting exposed to sexual content for the first time through pornography at this age leads to disturbance. They try to know more and more about it who educate themselves through experience without any knowledge of reality. When they go to schools, they develop instant attraction toward the opposite sex and look out for practical implications without bothering about its pros and cons [6]. Too much watching of such contents lead to such situation and in case the person is unable to get it in practice, he or she may resort to bad habits like drugs, or alcohol.

The content in pornography is purely meant for entertainment which is completely baseless and meaningless prioritizing on sexual gratification [7]. When it is watched by the young adolescent children, they will not be able to distinguish between a relationship or having sex which proves to be more harmful. It is an impact of watching too much porn content where the mind becomes conditioned to sex differ from a relationship. In this situation, the boys and girls try to get fictitious by imagining themselves in place of the actors thinking that this is the reality [8]. Sometimes when their expectations do not match and could not get aroused, they are victimized of body shaming, they cannot face their classmates due to humiliation.

9.3 Psychological Impact of Pornography

As discussed earlier, pornographic content is mainly prepared to generate sexual interest and satisfaction. There are many online users who have developed the habit of watching it on a regular basis. However, watching it without control excessively could lead to depression and rise in anxiety levels and sometimes sexual dysfunction. In the process they get addicted to watching porn which may lead to spending time alone causing rise in delinquent behavior [9]. They would want to spend less time with their family and get detached with those people who care about them. Though there have been research which were conducted on the psychological impact on both males and females, interestingly women who watched porn were found to be giving more value to their relationships with their partners where as it was found to be opposite in case of male respondents though sometimes findings could be both negative as well as positive with low relationship satisfaction [10].

9.4 Review of Literature (Existing Models)

As it is known widely, globalization has brought access to technology and people are more prone to using the latest versions. Cell phones and other communication devices like laptops have become very much available for individuals of all age groups due to which they get to access various sites with much ease. Few of the most viewed sites are the porn sites which generate interest among the users of all age groups. They consider watching porn videos to be one of the best sources of entertainment. Some of them get so much addicted to these porn videos that they initiate sexting, forcing their partners into having sex, kidnapping, and molesting. Due to revolution in technology, users are more used to get facilitated by inappropriate content, particularly nudity and pornographic visual content whether intentionally or unintentionally leading to a change in their behavior. This is one of the reasons why filters need to be incorporated which can churn out porn contents and those media which contain eroticism and violence. Various researches have been undertaken on this topic. Some of them are as follows: Studies have shown that early exposure to adult content could promote negative mental health and increase the intention to engage in sexual activities [11, 12]. In addition, unrealistic sexual beliefs may be developed in young pornography consumers. This poses a concern to society, especially to parents, which calls for a need for control measures to limit the exposure to pornographic images and videos as humans are visual beings. Automation of pornography recognition using deep learning techniques has greatly improved the efficiency of censorship by reducing the workforce of censorship editors. In 2015, a pornographic video classifier was designed by taking classifications of fine-tuned convolutional neural networks (CNN) such as AlexNet and GoogLeNet on video keyframes into account. Usage of a support vector machine (SVM) to classify features detected by CNN models was also applied to recognize obscene video frames [13]. Furthermore, Aldahoul et al. [2] tested the performance of pornography detection in cartoon videos by combining decisions of several fusion approaches that utilize CNN as a feature extractor and SVM as a classifier. A multi-level pornographic image classifier was implemented by using ResNet-50 and Mask R-CNN models such that images with low classification probabilities would be sent to the consequent stage while the rest would carry the first stage decisions. Usage of the You Only Look Once v3 (YOLOv3) object detector to focus on image patches containing humans has improved the classification performance of CNN on images with small-scale pornographic content [2].

There are various pornography detection models available on the internet which are the following:

9.4.1 Female Breast Class

Due to the sensitive nature of pornographic content, certain classes appear more frequently than others, as they are common in a wider range of pornographic scenarios. This is particularly noticeable in the over-representation of the "Female Breast" class, which is expected to remain imbalanced regardless of the volume of collected data [14]. To address this imbalance, we apply six standard data augmentation techniques to images containing the two most underrepresented classes, namely "Anal" and "Male Buttock". This augmentation helps increase the number of usable examples for these classes. It's important to note that while data augmentation is applied to specific images with underrepresented classes, these images may also contain instances of other classes, unintentionally increasing the number of overrepresented classes.

For our main architecture, we chose the YOLOv3 real-time object detector because it can be trained quickly and utilizes the entire image as additional context for boundary prediction, resulting in fewer false positives compared to its closest competitor, the faster-RCNN. We selected an image input size of 416 × 416, which strikes a balance between minimizing computational requirements per inference and maximizing model precision. Redmon et al. reported that at this size, YOLOv3 achieved a mean Average Precision 50 (mAP-50) score of 55.3 with an inference time of 29 ms, compared to 57.5 obtained by RetinaNet-101-800 in 198 ms [15].

To enhance the generalization of the Sexual Object Detection (SOD) model and reduce false positives from non-explicit images, our training process is divided into two stages with separate sets of weights. In the first stage, we load pretrained ImageNet weights and freeze the image feature extraction part of the YOLOv3 model. This enables the object detection portion of the model to be trained using general image features learned by the DarkNet-53 base model on the ImageNet dataset. This reduces the number of parameters to be adjusted, speeding up the training process. We use a batch size of 64, implement early stopping with a patience of ten, reduce the learning rate on a plateau by a factor of 0.1 with a patience of three, employ the Adam optimizer with an initial learning rate of 10^{-4}, and train for a maximum of 51 epochs. Once the first stage concludes, the weights yielding the best validation loss are retained.

In the second stage, the current model is unfrozen, and end-to-end training continues using a batch size of six while keeping the other parameters unchanged. Again, the weights yielding the best validation loss are retained. The model generates multiple candidate boxes per image, which are filtered using a Non-Max Suppression (NMS) algorithm. Boxes with confidence scores below a specified threshold are removed, and the remaining boxes are grouped by class. In cases where multiple boxes of the same class overlap, the box with the highest confidence is retained. While most classes can overlap, the "Female Genitalia" class is split into two sub-classes representing two distinct appearances, and therefore, the NMS algorithm is modified to consider and suppress overlapping instances of this class pair jointly.

9.4.2 Multi-confidence Thresholding

Multi-confidence thresholding could prove to be a more effective method of detecting porn content. In this technique data augmentation is usually performed which improvises the presentability of that class which is not represented in a proper way. It is assumed that some of the models were exposed more in some classes and less in the other, it can be inferred that the confidence range is expected to differ in different classes. At the same time, if single confidence threshold is created for all the classes, it may initiate biasism toward some number of classes. In order to overcome this, in this section we describe a methodology that attempts to extract any class confidence biases exhibited. In doing so, we aim to obtain a vector of optimal class-specific confidence thresholds that will replace the single confidence threshold set in the original inference. A small holdout set of 191 pornographic images was set aside when creating the main dataset [16]. To represent how the model suppresses false positives in non-pornographic images, we add 50 benign images to this holdout set. We proceed to infer the class labels for these 300 images at a very low confidence threshold of 0.05 and collect a set of potential predictions. These potential predictions are compared to their corresponding annotations and marked as either true positives, false positives, false negatives, or misclassifications which illustrates the distribution of false positives against true positives for two classes obtained from the test set. Using this illustration it becomes evident that the optimal confidence threshold for each class is different and can also depend on the user's preference towards prioritizing either the precision or recall of the model.

9.4.3 Overview of the Trained Sexual Content Classification Model

The sexual content classification model utilizes the YOLOv3 architecture, consisting of three main blocks. The first block is DarkNet53, a custom CNN architecture serving as the backbone to extract image features. Object detection is performed using three different sizes of feature maps and additional CNN layers [17]. The model generates candidate boxes, which are then processed through a non-max suppression stage to obtain the final result.

Normally, a classification task like this would require a dedicated CNN network. Even a lightweight architecture like MobileNetV2 would need millions of additional parameters to extract sufficient image features for successful classification. However, instead of training a new classifier from scratch, it is hypothesized that the SOD (Salient Object Detection) model has already extracted enough information from the image for the additional classification task [18]. To test this hypothesis, a standard classification head is employed, consisting of two fully connected layers followed by a 19-way Softmax layer. Four different topologies were implemented, with only the newly added portions modified during training to preserve SOD performance.

The image feature extraction layer's top layer is used similarly to traditional CNN classifiers, while Node 2 and 4 mimic the YOLOv3 multi-scale detection method by concatenating the three image feature levels. Additional residual blocks are included to extract further features driven by the classification loss. However, this incurs a significant increase in the number of parameters required.

In the second component of the model, object detection features are used instead of image features, as the local features of explicit objects may strongly correlate with pornographic content. During the initial training phase, the original YOLOv3 architecture with SOD weights is employed. To preserve the original object detection performance, all model layers are "frozen", ensuring the parameter values remain unchanged during training. After adding the new classification node, its output layer is assigned a categorical loss function, while the three original object detection outputs are assigned "None" values, indicating no loss calculation for those outputs during training.

The model training process involves a validation split of 0.15, a batch size of 64, and the use of the "Adam" optimizer. Early stopping is implemented, ending the training process if the validation loss does not improve after 16 epochs. Additionally, the learning rate is reduced by a factor of 0.1 if the validation loss does not improve after 5 epochs [17].

9.4.4 Probabilistic Latent Semantic Analysis

Probabilistic Latent Semantic Analysis (PLSA) is a statistical model that differs from the standard Latent Semantic Analysis (LSA). PLSA offers several advantages, including the ability to utilize standard statistical methods for model fitting, complexity control, and model selection. For example, the performance of a PLSA model can be evaluated by assessing its predictive accuracy through techniques like cross-validation. PLSA employs a latent topic layer that connects documents and words, representing each document as a mixture of concepts weighted by the probability P(zjd), and each word expressing a topic with probability P(wjz) [19]. The structure of PLSA resembles a statistical aspect model.

In the domain of nudity detection using skin information, various techniques have been explored. Fleck et al. proposed a two-step content-based retrieval strategy for identifying images with naked people based on skin regions. Skin pixels are identified by applying thresholds to the intensity, hue, and saturation values of each pixel. However, this approach is susceptible to scale and saturation issues, leading to false positives. Geometrical analysis of the skin regions also suffers from challenges such as occlusion, close-ups, or failures in the skin detector. These limitations result in lower precision and recall measures compared to newer methods. Jones and Rehg focused on color information, constructing skin-based statistical models by computing histograms of skin and non-skin colors. A likelihood ratio approach is used to label an RGB value as skin if it exceeds a certain threshold. Feature vectors are then created, including the number of detected skin pixels and the average confidence of the detected skin, and a decision tree classifier is employed for decision-making [20].

9.5 Comparative Analysis

In the previously discussed model, the architecture essentially involved deep learning and neural networks to deduce and filter pornographic content. The model however had their own limitations and benefits which could be taken into account while designing a hypothetical model that specifically detects female exploitation and gender violence. Thus, this section explores in this domain to compare the discussed models and conclude a hypothetical model to filter pornographic material catered to gender violence [21].

For instance, in the first model where the underrepresented classes are considered for filtering Sexual Object Detection (SOD). Among these under-represented classes, there are "Anal" and "male buttocks" to balance the gap between under-represented and over-represented categories. In order to detect potential female exploitation, we hypothesize to consider possible keywords that specifically relate to abuse of women or gender violation such as "forced", "rape", "slave", etc. Providing these keywords to the model could help us classify videos that are unethically posted as pornographic content on the internet. Since these words will filter out content particular to violence category, it could speed up investigation of possible crimes.

During the multi-confidence thresholding, the dataset created could include pornographic images hinting at suspicious behavior like convicted cybercrime images, videos that could help detect similar patterns during detection. Thus the confidence threshold could be altered to ensure precision in detecting potential sexual abuse in women. Now, using this adjusted SOD model, we could create a classification model as specified earlier. The layers could be arranged in accordance with the dataset and a model could be trained to achieve our desired result.

Finally, during the Probabilistic Latent Semantic Analysis (PLSA), we can define our data model that would be efficient enough to achieve our goal of detecting and eradicating female exploitation through porn. PLSA can easily determine whether or not our new model will be suitable to aid our investigation. It could measure units of performance and other factors that are crucial for the purpose. Thus, following through these steps as well as changing them to match our interests could result in an effective model.

9.6 Conclusion

Thus the evaluation of our techniques shows that the association of Deep Learning with the combined use of static and motion information considerably improves pornography detection. The paper proposed a new methodology to compare and also analyze various detection models which use Deep Learning technique, after which the best model was inferred.

References

1. Abdulla, W. (2017). *Mask R-CNN for object detection and instance segmentation on Keras and TensorFlow*. https://github.com/matterport/Mask_RCNN
2. AlDahoul, N., Karim, H. A., Abdullah, M. H. L., Fauzi, M. F. A., Wazir, A. S. B., Mansor, S., and, & See, J. (2020). Transfer detection of YOLO to focus CNN's attention on nude regions for adult content detection. *Symmetry, 13*(1), 26. https://doi.org/10.3390/sym13010026
3. Avila, S., Thome, N., Cord, M., Valle, E., & Araújo, A. A. (2013). Pooling in image representation: The visual codeword point of view. *Computer Vision and Image Understanding, 117*(5), 453–465. https://doi.org/10.1016/j.cviu.2012.09.007
4. Bernardi, R., Cakici, R., Elliott, D., Erdem, A., Erdem, E., IkizlerCinbis, N., Keller, F., Muscat, A., & Plank, B. (2016). Automatic description generation from images: A survey of models, datasets, and evaluation measures. *Journal of Artificial Intelligence Research, 55*, 409–442. https://doi.org/10.1613/jair.4900
5. Broadhurst, R. (2019). Child sex abuse images and exploitation materials. In *The human factor of cybercrime* (pp. 310–336). Routledge. https://doi.org/10.4324/9780429460593_14
6. Devlin, J., Cheng, H., Fang, H., Gupta, S., Deng, L., He, X., Zweig, G., & Mitchell, M. (2015). Language models for image captioning: The quirks and what works. (2015). arXiv: http://arxiv.org/abs/1505.01809v3 [cs.CL].
7. Hamian, M. H., Beikmohammadi, A., Ahmadi, A., & Nasersharif, B. (2021). Semantic segmentation of autonomous driving images by the combination of deep learning and classical segmentation. In *2021 26th International Computer Conference, Computer Society of Iran (CSICC)* (pp. 1–6). https://doi.org/10.1109/CSICC52343.2021.9420573
8. Hamid, M. S., Manap, N. F. A., Hamzah, R. A., & Kadmin, A. F. (2020). Stereo matching algorithm based on deep learning: A survey. *Journal of King Saud University - Computer and Information Sciences*. https://doi.org/10.1016/j.jksuci.2020.08.011
9. K. He, X. Zhang, S. Ren, and J. Sun, 2016 "Deep residual learning for image recognition," in 2016 IEEE conference on Computer Vision and Pattern Recognition (CVPR), pp. 770–778.
10. He, K., Gkioxari, G., Dollar, P., & Girshick, R. (2020). Mask r-cnn. In *IEEE International Conference on Computer Vision (ICCV)* (pp. 2980–2988).
11. Zhang J., & Jemmott J. B. (2015). Unintentional exposure to online sexual content and sexual behavior intentions among college students in China. *Asia Pacific Journal of Public Health. 27*(5):561–571. https://doi.org/10.1177/1010539514562446
12. Hodosh, M., Young, P., & Hockenmaier, J. (2013). Framing image description as a ranking task: Data, models and evaluation metrics. *Journal of Artificial Intelligence Research, 47*, 853–899. https://doi.org/10.1613/jair.3994
13. Lizhi Huang and Xunyi Ren. (2018). Erotic image recognition method of bagging integrated convolutional neural network. In *Proceedings of the 2nd international conference on computer science and application engineering (CSAE '18)*. ACM, Article 107, 7 pages. https://doi.org/10.1145/3207677.3277990.
14. Lin, T.-Y., Dollar, P., Girshick, R., He, K., Hariharan, B., & Belongie, S. (2017). Feature pyramid networks for object detection. In *IEEE Conference on Computer Vision and Pattern Recognition (CVPR)* (pp. 936–944).
15. Nguyen, Q. H., Nguyen, K. N. K., Tran, H. L., Nguyen, T. T., Phan, D. D., & Vu, D. L. (2020). Multi-level detector for pornographic content using cnn models. In *RIVF International Conference on Computing and Communication Technologies (RIVF)* (pp. 1–5).
16. Nugroho, H. A., Hardiyanto, D., & Adji, T. B. (2016). Nipple detection to identify negative content on digital images. In *2016 International Seminar on Intelligent Technology and Its Applications (ISITIA)* (pp. 43–48).
17. Redmon, J., & Farhadi, A. (2017). Yolo9000: Better, faster, stronger. In *2017 IEEE Conference on Computer Vision and Pattern Recognition (CVPR)* (pp. 6517–6525).
18. Tabone, A., Bonnici, A., Cristina, S., Farrugia, R. A., & Camilleri, K. P. (2016). Private body part detection using deep learning. In *ICPRAM* (pp. 205–211).

19. Tian, C., Zhang, X., Wei, W., & Gao, X. (2018). Color pornographic image detection based on color-saliency preserved mixture deformable part model. *Multimedia Tools and Applications, 77*(6), 6629–6645.
20. Tran, H. L., Nguyen, Q. H., Phan, D. D., Nguyen, T. T., Vu, D. L., et al. (2020). Additional learning on object detection: A novel approach in pornography classification. In *International conference on future data and security engineering* (pp. 311–324). Springer.
21. Wang, Y., Jin, X., & Tan, X. (2016). Pornographic image recognition by strongly-supervised deep multiple instance learning. In *2016 IEEE International Conference on Image Processing (ICIP)* (pp. 4418–4422).

Chapter 10
Framing the Landscape of Technological Enhancements: Artificial Intelligence, Gender Issues, and Ethical Dilemmas

Subhankar Dutta

10.1 Introduction

The recent decades have witnessed a new emergence of digital technology, taking different shades and shapes now and then. Especially the emergence of AI (Artificial Intelligence) and ML (Machine Learning) has revolutionised the STEM fields considerably. As AI unfolds, new questions are taking shape around ethical concerns, gender biases, job possibilities, and other emerging uncertainties looming around the developing technology. Along with the posthumanist concerns around the worldwide environmental crisis, the present generation of scientists, engineers, and thinkers are trying to make sense of a new world created by the human, for the human, but may not be 'of the human'. Therefore, the concern around AI is less about what it is but more about what it can be, and what it ought to be. AI or artificial intelligence refers to the capability of a computer to classify, analyse, and draw predictions from an extensive data set using backhand patterns called algorithms. On the other hand, the ability of an AI system to learn and improve automatically from the experience and results of the earlier data sets is broadly called machine learning. Thus, the primary establishment around AI technology aims at mimicking the human brain and decision-making system to its utmost capacity and beyond. From the automated voice recognition system in our smartphones to the more extensive satellite image processing, AI technology is showing us a future that has yet to explore fully. However, the discussion around AI and the prediction of such computer technology is a relatively older phenomenon [11]. The recent successful application and implication have become new concerns across sectarian boundaries.

S. Dutta (✉)
Department of Humanities and Social Sciences, Indian Institute of Technology Bombay, Mumbai, India
e-mail: subhankardutta@iitb.ac.in

D. Mishra et al. (eds.), *Communication Technology and Gender Violence*, Signals and Communication Technology, https://doi.org/10.1007/978-3-031-45237-6_10

Biomedical engineering, medicine, robotics, and the entertainment industry rely hugely on artificial intelligence to solve life-like and larger-than-life problems. While AI technology can be used to make our life more accessible than before, AI failures, which are primarily human failures, raise concerns about the significant disasters that can occur. It is not only a concern about technological failure, but it has broader social, cultural, and material implications which are long-lasting. While having a self-driving car or finding the perfect match over a dating application can look fascinating, there are underlying concerns and problems in making AI technology that can reaffirm existing social biases and reframe the existing gender violence on the web. Therefore, AI design, development, deployment, and governance are significant aspects to ensure that 'AI is for good'. Not only this, the rapid development of AI technology and the unpredictable nature of its development have also raised questions on the job market, human labour, gender biases, to other more significant ethical concerns.

The present paper engages with one such aspect of AI and machine learning that has major implications for our society, culture, and technological development. The increased advancement of AI in creating opinions and behaviour in everyday life is raising concerns about the design of the technology, which can undo the significant development around gender equality. The larger social structure and cultural ecosystem have long-established gender norms that have a complex relationship with personal experience and communal harmony. Being predominantly a male arena, the AI data field is laden with stereotypical concepts that have a broader risk of manipulating the machine incorrectly and eventually creating an affecting bias that can create loopholes for gender-based violence. Gender ideologies are embedded in language, culture, behaviour, and social conduct. Therefore, it demands a thorough engagement to ensure a safe space for technological enhancement. The paper delves into a few significant gender concerns that effectively question the visible and invisible gender violence formulated by AI technology and machine learning. This new global endemic of stress around technological advancements also brings concerns around post-work societies, post-work economics, and post-work politics. As the liberal story is falling [9], how will ideas of free will, governance, and equality be configured in the coming time? From the creation of the faith-machine google to the new bio-techno-medical development, algorithms will repeatedly be faulty. Therefore, who will be the thinking human and the decision-maker in this big data world? The paper broadly re-imagines these broader questions, taking gender violence in AI technology as a primary concern. The challenges and possibilities of the technological development around AI and machine learning are discussed with an analytical lens positioned on the gender-related violence of the present and the future to unfold.

10.2 Representation, Development, and Fault Lines: The Importance of Looking into Gender Violence in Relation to AI

All that glitters is not gold. The concern around technological enhancement has a similar dimension and correlation, which is not new. Technological development has always been a two-sided weapon, often used for unfair means and used destructively also. Therefore, the development logic of any technology often remains under a dark shadow of ethical uncertainty, and more significantly, human rights concerns. Regarding gender violence caused by/in AI and machine learning, it is more complex, subtle, ingrained, and very implicitly operational. So, gender as a category under the larger spectrum of AI needs special mention and a critical eye. AI technology and gender biases operate in a multilayered fashion and have different spectrums where it is functional, sometimes even before data collection. The purpose is to build a system where the biases of today do not impact the technologies of tomorrow. Therefore, it is essential to comprehend what makes gender such a crucial concern within the recent development of AI technology and machine learning. What are the implications of these new avenues that get into our daily life with much more automation and robotisation? How the technological enhancements facilitate gender-based violence?

The primary benchmark in the implication of these new technological devices has been performance. The accuracy and performance dynamics have inherent issues as it often overlooks the gender differences in seeking quick accuracy and conclusiveness. AI technology and, most notably, the future with it are going to be complex. The desire to simplify technology often comes with rejecting small biases, which can have enormous implications in the coming time. Therefore, it is more important to look at AI technology as something reflecting gender biases not only fed outwardly but also inherent within our societal system. It is a focused area that needs rectification and a crucial sector that can reflect the fault lines in the so-long development of gender equality in our society. It is both a reflection and a mirror to be taken seriously.

Feminist studies have long engaged with the concerns around gender biases in literature and society, where females are portrayed as inferior beings who are passive and irrational [17]. In the twentieth century, feminist scholars like Judith Butler propounded the ideas about biases inherent within the very language usage, perpetuating a typical gender ideology in society [2]. These ideologies are embedded in the text resources and, many a time, in societies. These eventually result in machine learning algorithms creating banal concepts about gender. The prevalent power dynamics of the society also contribute to the effective biases that often remain below the carpet. The long-drawn written medium and human consciousness are challenged by a machine interface which is not intuitional or context-driven but more data-driven. Ordering male counterparts often before females in conventional naming patterns demonstrates a social bias in practice that machines can replicate. Son-daughter, husband-wife, male-female to even master-servant, doctor-nurse are all associated with a power dynamism working much before AI came to the fore [19]. With the massive implementation of AI into our day-to-day practices, as it will grow to another extent in the coming decades: such minute biases will have huge data bias and become facilitator of gender-related violence. Many small to significant biases can be effectively operational in our societies that are getting carried forward to AI development through these narrow paths. For example, androcentric terms such as 'he', 'him', 'man', and 'mankind' are often used to refer to both men and women. Though this might sound like a minor issue, its application in machine learning can have a larger implication where the machine can act biased towards the feminine gender [14]. Based on this problem, many professions might be referred to by the machine as exclusive to the male gender. Eventually, individual discrimination will contribute to collective biases in matters of group identification. These outdated biases, enormously used in our daily life and textual applications, will also create erroneous source data and machine knowledge. The datafication of the technological industry is a recent phenomenon after the beginning of data science. However, the data gathering in the system is a swift and rapid process. AI-based systems that we use daily can accumulate an enormous amount of data and have the capacity to process it in seconds [15]. So, AI and related technology have a profound information terrain that will affect the world order in numerous ways. The feminine gender, because of their historical baggage of oppression, discrimination, and transitions, will be one of the most impacted groups that have to go through a massive change. The potential and the pitfalls will impact women's lives and conducts in a technologically changing society. Understanding the patterns of this change that emerging AI technology and machine learning are bringing will lay bare the power of adaptability that the technology entails and its potential. If sexism or gender biases are embedded within the system, the development dynamics of the field of artificial intelligence will exhibit the same sexist or gender-biased behaviour.

Before looking into the specificities of AI technology and its implication in gender violence, it is important to see the link between technology in general and technology-facilitated gender violence. Gender-based violence is a global concern where individuals and groups who identify beyond the normative patterns of gender identities are subject to various forms of violence. As in recent decades, we have increasingly become more digitally mediated, gender violence also shifted to a digital spectrum, making it more complex. Starting from individual bullying over social media to larger forms of digital harassment, digital technology has brought the accessibility of violence to a more anonymous and targeted form. As it provided convenient tools for lifestyle, it also made violence easier in matters of execution. Unlike physical violence, which requires the presence of people, digital violence has gone beyond national and continental boundaries and has become a global concern. Going beyond the geographical boundaries, this digital form of violence put the process of law enforcement and policy-making into a problematic state. The state power and the digital protection laws, peculiar to different countries, are also providing loopholes to the creator of violence. Copied information, worldwide circulation, and the absence of source content at times also retain the perpetual risk of future violence. Whereas the judicial system, policy making, law and order are trying to minimise the gender-based violence actualised over digital space, the concern around AI technology adds new dynamics to the anxiety. The homogeneity of the AI field, with less female representation and undefined policy, is evolving into a gender-biased technological space, thereby giving shape to various forms of gender-based violence. As per the *World Economic Forum* report on the development of AI technology, only 22% of AI professionals globally are female, compared to 78% who are male [27]. This huge rift between the two genders has effectively caused biased and limited representation and sensitivity in technology enhancement resulting in half-fed machines with limited data. According to *Bloomberg*, a leading software company, among eight large tech companies, only 20% of technical roles are filled by women employees [3]. Globally, women account for only 25% of workers in science, technology, engineering and mathematics (STEM) but only make up 9% of leaders in those fields, according to the *Boston Consulting Group* [13]. Frida Polli, chief executive officer of *Pymetrics*, referred to the danger of such gaps in the male-female ratio and said, "Can you imagine if all the toddlers in the world were raised by 20-year-old men? That's what our AI looks like today. It's being built by a very homogenous group" (as quoted in [1, para. 5]).

However, beyond the apparent gender concerns of representation, it is also crucial to see that AI might fall into a gender binary setup. The violence on digital platforms is no more limited to a binary setup of gender, but it also includes non-binary and gender non-conforming individuals. The various forms of violence over the digital spectrum, like harassing digital communication, impersonation, trafficking, and disclosure of personal information and data, give rise to a new set of anxieties coupled with AI technology, which is more automated and has minimal human control. Therefore, the technology should also take into account the non-binary and transgender expressions very much prevalent in our modern social systems. We need a trans-inclusive AI system that can undo many effective practices of gender violence in society. Gender identity, essentially an individual choice, will contradict with the automation that machine learning is looking for based on a large data set [10]. Bodies, outward expressions, facial structure, and automated image processing techniques need to have a more inclusive and well-thought design so that the system remains sensible for transgender, genderqueer, and self-identified genders. Broadly, there are two significant aspects to AI and machine learning concerns. Firstly, to access and see the essential link between the data fed to the AI system and the potential implication of the same on policy-making. Secondly, to see the inherent biases and shortcomings within the AI technology itself. The most challenging task in this regard is to demonstrate and configure how abstract concepts like gender can be operationalised into measurable units that can be computationally identified. This building of connection between theoretical concerns and the application or feature extraction is crucial in effectively addressing gender violence. For this, it also needs to be ensured that the AI is developed for social good, having a clear-cut framework and definition of what this ‘good’ stands for. Questions around authority, implementation, and, most importantly, intentions demand due attention to have a positive outcome in socially relevant spheres.

10.3 AI for Social Good: Gender Violence and Technology

Over the past few decades, scholars and social scientists have significantly dealt with the risk and concerns of AI development, which needs a more sustainable setup and operation. The policymakers and the governments have also started a considerable effort to ensure risk management and fewer data infringements. An array of efforts and initiatives are already underway; for example, the UN has developed AI-based responses to help meet the UN’s Sustainable Development Goals (SDGs) by 2030 [6, 23]. The global pandemic and the climate crisis have also laid bare the all-encompassing concerns about the global technological crisis and possibilities in answering life-like and larger-than-life problems. So, in practice and principle, the focus must be on sustainable development and serving specific purposes. A series of empirically and ethically grounded concerns can ensure that AI is not being used for evil purposes. Gender, one of its crucial aspects, needs to be examined from this social responsibility perspective. The crucial challenges in making AI for social good are also related to the rapid growth of the private firms and bodies which are hugely developing and experimenting with such technologies. These private bodies occupy a dominant position in technology development, implementation, and distribution [5]. The successful implementation of the technology in products and the circulation of these for-profit organisations in the market are vast and very rapid, often raising the question of whose good it serves. Due to this accessibility and anonymity that the digital technology provides, there are numerous forms of harm caused by digital perpetrators in general and AI makers in particular. The maximum violence over digital platforms comes from the huge mushrooming of social media platforms where anonymous comments, public accessibility, and the digital algorithm point out a trail of thought to a single directional crowd. People with similar thoughts and intentions of digital bullying, perform in a digitally mediated community setup. It is actualised through non-gender sensitive digital media platforms where gender sensitivity is subjective, and the digital spectrum provides that platform. The sociability of digital space and the literacy of gender sensitivity on a digital platform are still away from practice. This leads to discriminatory comments on gender, discouraging free thoughts of women activists on gender terms, and other subtle forms of bullying leading to psychological trauma, reputational hazards, and also severe professional consequences. It reinforces the inequality and discrimination that women and other transgender people are already subjected to, curtailing their basic human rights of expressing themselves. It qualifies a patriarchic order and gives shapes to other forms of intersectional politics, social hierarchies, and a looming digital divide. A report by Battered Women’s Support Services in Vancouver, British Columbia, points out that 64% of the surveyed women reported psychological trauma and mental anxiety caused by digital harassment [26]. In the Indian context, as per a report, 28% of women out of 326 women are anxious and suffering from depression due to digital violence [8]. It effectively silences the voices of women and other non-conforming genders over the digital space, reflecting an unequal spectrum in the beyond-digital social setup. The fundamental reason behind such digital violence is the lack of privacy and safety measures in the digital world. The inclusion of AI technology should ideally ensure a safer and digitally end-to-end encrypted communication world. The complexity of the technology and the euphemistic nature of the technology thus become a significant concern in the discussion of digital technology and violence. Be it hacking into a digital device of a woman, filming or sharing private information over a public medium, or non-consensual digital publicity – all points toward the need for a conscious development of future technologies that are more gender-sensitive and address all these prominent concerns around gender-based violence.

The ongoing criticism of gender violence and the implementation or development of AI, therefore, demands a feminist epistemology to be engaged and concerned with making a sensible atmosphere where the social concerns around gender can be taken care of by the system. It needs a textually, historically, and linguistically situated and contextualised system which can undo and not redo the dogmatic aspects like memory bias, faulty face recognition, euphemistic intelligence, and other gender concerns. Researchers, scientists, and authorities must rely on an interdisciplinary realm where social issues, AI, and other intersections can interact to lay bare the complex issues that might arise later. To ensure a better technological space where gender violence is minimal and technological development helps in having a more egalitarian digital space, we must ensure that AI is used for social good. The foremost thing to keep in concern is the coherence among various components that make the design and deployment of the technology successful. In this regard, the idea of trust, therefore, has become very crucial in the technological development of AI. However, how do we define trust in this respect? It is not about the performing nature of the technology or the accuracy but more about the development of a technology, sensitive to critical social causes: a system that does not discriminate, a system in which we can understand how it works—a safe and non-manipulative system at hand. The system should have one or multiple layers of empirical testing where the falsifiability of the system can go through. An operational framework should be there, without which the system could not or should not work. As the technological system needs a direct application to test the trustworthiness of the system and its authenticity, "the map of testing would simply equate to the territory of deployment" [7, p. 1776]. It is essential to identify and work through the faulty layers of the technology even before its deployment in the market. Once it is out there, we can only know whether the technology is faulty or not, but we cannot fix it largely. Floridi suggested an 'incremental deployment system' where the hazardous effects of the system can be tested beforehand by the developers by having falsifiable assumptions and application of the same in the protected context [7]. As per the requirement of the system, the testing set should be well large enough to ensure clarity on sustainability and trustworthiness. It should also have a systemic level where the whole application can be stopped at any given point whenever any malicious component gets detected. The simulation set for testing should essentially look at replicating the real world without biases like gender. For example, the development of the game with suggestive automation of soldiers and car drivers primarily creates a simulative atmosphere where a male driver or male soldier is driving or leading a team of soldiers, respectively. The testing of the game within a small simulative set of the natural world might stand out as a reliable set. However, it co-creates a social stigma that has been prevalent in society for a long time. The assumption model of the system in the real world can bring several hazards in matters of applicability and deployment. Therefore, routine processing of the outputs and constant update on the programme's progress is essential in keeping the growth of the bias minimal and cutting short the number of violence. As the future of technological development will rely on the futuristic visionary of the AI system a lot, the reliability of the result is way more important than the efficacy of the system at present. Data infringement and unreliability of the data set is one of the other crucial concerns that need to be taken critically to have proper AI development. For example, the medical engineering sector, which is relying a lot now on the computerised detection system and image processing method of disease detection, can be hugely disruptive if the data set is not cohesive and coherent with the system demand. Geographical location, local concerns, the gender ratio of that particular data set, and racial concerns (even in the Indian context, caste differences can be seen as significant) come as challenges that can maximise the bias in accuracy and transparency.

Therefore, thorough contextualisation of the data set and contextualised intervention are necessary for socially and technically reliable AI. The intervention in the data set is a two-way affair. On the one hand, the training data can be manipulative and incomplete in nature. Simultaneously gathering data from the users can bring more accuracy and correct the earlier biases inherent within the system. This is more concerning machine learning, where the system takes the user as a co-developer in the data management and production process. Few autonomous decision-making abilities are ascribed to the user, eventually making the master data set more potent and accurate. In this regard, it is more important to have a proper levelling of the disruption as it can also bring new biases to the system with unnecessary data gatherings, which is not required at all. Preferences, choices, and selections at all levels regarding these need to be in sync, which can ultimately contribute to the goal of the system and the learning process of the machine [16]. There is inevitable interdependence among all these factors that collectively make the system analyse and learn from the data. The user and the developer need to have a more concrete relationship based on ethical considerations where technology is not only driven by a particular purpose but also the purpose can be well explained and ethically appropriated. The design of the algorithm should be responsive engineering where the system development method is contextualised and well explained to the receivers. To Watcher, this should include information about the general functionality of the system and the logic that is being used in its making, and explanations for every decision taken within the system [24].

In pursuing bias-free AI technology, among several other essential factors, the privacy concern is significant as it entails challenges of earlier gender norms and systemic faults of our society. Dignity, social cohesion, and human rights are concerns that feminists and social activists were already been fighting for decades. Also, the technological development before the oncoming of AI already had many biases regarding data privacy [20]. So, in making AI for social good, privacy stands out to be a genuine concern. It is the space where policy-making and governance have significant roles to play. At the collective level, there have been rules and regulations from the concerned authority, like the General Data Protection Law in Europe and the Act on Protection of Personal Information in Japan. Also, in India, judicial decisions jurisdictions have been taken by the central government to ensure data protection [18]. Beyond this systemic level, there needs to be a more elaborate policy on AI development and deployment to ensure that personal information and human right to privacy are not hampered at the fundamental and operational levels of AI technology. Ensuring privacy consent is the foremost requirement that the system should have from the users, elaborately. On a very granular level, the users should have access to change the data privacy pertaining to their personal information and choices. J. Kaye talks of this as the *dynamic consent* system more concerning the medical sector where the patient can simultaneously co-create and monitor the privacy data sets to ensure the maximum practice of human rights in collecting personal details [12]. A similar example can be found in Wang and Kosinski's essay on deep neural networks, which talks about the randomisation of the images taken from internet sources to detect the sexuality of the user based on a minimal number of data sets [25]. This breach of privacy, where the data is used without any consent of the source, often leads to misrepresentation, digital-violence, and privacy concerns of larger order.

So, the socially significant data usages in AI technology have to be very cautious and conscious effort based on ethical considerations, subject to regulations, and sensitive to situational concerns. A more ethically rooted purpose needs to be the driving force in developing the technology in the coming time. Both in the larger implementation as well as at the foundational stage, an intrinsic balance is necessary by closely identifying the small factors legible for creating affective biases. More than protection, contextualisation and effective intervention turn out to be important in eliminating biases. In a commercially driven construction of AI technology, concerns like gender biases and ethical dilemmas cannot be answered or discussed in the abstract. As technology has a concrete foundation for its functions, the question should be at the fundamental of those foundations. This is not to eliminate the essential functionalities of the technology but to bring systemic balance, conscious deployment, and measurable fault lines that can be addressed in the process of its development. At this juncture, AI stands out to be a double-edged weapon. On the one hand, algorithms can magnify the existing biases in gender because of not having a sensitive focus on conflicting issues. On the other hand, it can also very effectively mitigate the biases by creating a inclusive digital space that can widen rapidly. The time not only demands a gender-neutral AI system, but more importantly, it looks for a gender-sensitive technology that can make human-computer interaction more vivid and humane. The power of awareness of sex and gender differentiation in AI can lead to better medical treatment, robotic facilities, healthcare, and minimise gender-based violence. To do that, awareness and education will be two vital components which will process the AI into a practical social benefactor. Data authenticity is not something essentially fed but more realised and principled on ethical grounds. Therefore having a proper AI principle focused on gender dynamics is essential to ensure that there are feminist internet principles which can fill the existing gaps and address the existing challenges at the systemic level.

10.4 The Future of Technology and Gender Equality: A Path Forward

Gender violence over the digital spectrum is a growing international concern affecting regions. It demands a thorough gaze and discussion at all levels to ensure a safer digital space and technological development. The first step is identifying loopholes where digitally mediated forms of violence are practised. Harassment, defamation, misrepresentation, privacy issues, hate speech, and other active forms of digital violence need to be marked. Secondly, pointing out the groups and sections of the populace, who are most affected, is essential to categorically make inclusive digital development as per the needs and concerns of specific communities. It is not to look for digital exclusivity but to have more inclusive units that provide a more egalitarian digital space. Along with the responsibility that the techno industry entails in ensuring this, the private sector, governments, inter-governmental organisations, civil society and academia have significant functions in making digital technology more reliable and bringing it within the larger human developmental spectrum. To ensure that digital policies are the foremost concern, organisations have to fix and rely on the development of technology which treats all gender equally and equitably. Though in many organisations, ethical principles have been documented from a feminist viewpoint, what lacks is a cohesive and guided principle which can collectively answer to all the violence or bias discussed earlier, especially concerning gender.

It is essential to categorically ponder a few such principles and see how effectively they can be implemented at the systemic level to eliminate or cut short the violence. It is not at the recommended level but more in the operational stage; these principles must be implemented. As the concern here is ethical, defining what these ethical considerations stand for is the most critical question. There always remains a gap between the suggestive principles, fairness recommendations, and demand for transparency with the awareness of structural development. The paper is trying to point out that we need to shift the narrative that AI as a technological system remains only within the constraints of the tech world. The framing of AI is a much more humanitarian concern than a robust approach to technological growth. This is the juncture where humanities and social sciences have a significant say in the construction of AI and for which awareness, education, and development of the right skill are essential. As the UNESCO report notes, "Global engineering education today is largely focused on scientific and technological courses that are not intrinsically related to the analysis of human values nor are overtly designed to positively increase human and environmental wellbeing" ([21, p. 19]).

Therefore, at the very fundamental level, the principles should be grounded on the purpose of education where training for ethically aligned AI technology is an integral part of technology-in-design. A more interdisciplinary communication can be implemented by having organisational principles where similar importance is provided to engineers, gender equality advocates, feminist scholars, and other practitioners. This educational pursuit will follow a more inclusive space within the STEM field where more girls and women are getting into it. There are definite shortcomings and barriers within the technological ecosystem preventing females from getting in and retaining themselves within a field like AI. For example, UNESCO's *Cracking the Code and I'd Blush if I Could* report categorically talks about the discrimination prevalent within the STEM field when it comes to gender equality [22]. Proper coordination between the top-down approach and the bottom-up approach is required to see the possibilities of tomorrow with more inclusive technology.

At the administrative level, the government needs to have strong policies that prioritise the development of technologies that respond to the specificities of women and contribute to the more significant cause of AI for social benefit. Through legislation and regulation, there should be multiple reviews processing of the product or the software before the general mass, or any third-party users can use it. The budgeting, institutional funding, institutional courses, and also training and scholarship should become gender-responsive. As a subsidiary of this effort, there should be a specific drive for recruiting female engineers and social scientists experts into the various governmental sectors, gradually working with emerging technology or implementing the same. Building government data banks and new institutions for training purposes, encouraging collaborating projects where the technicians can work with the people who are most affected by algorithm inequalities and having more women from the marginalised communities as employees and users can be effective forward movements from the government in ensuring fair techno development. Whereas AI still lacks global connectivity as per the rules and regulations, the product has a more visible circulation across the globe due to the worldwide internet network. Different countries, with contextual analysis, came up with different rules, principles, and regulations on using the technology. However, there should be a coherent global effort to consider the case of AI minutely and look for suitable and applicable principles. Along with the respective government, there needs to be a considerable effort from organisations like UNESCO to fight for larger social needs and equality.

A corporate governance principle, where the models and mechanism of the workforce are equally distributed and ethically considers the issues of parenting to the different gender, is crucial. It is not only the technology that needs to have a more gender-balanced future but also the whole technology industry needs to implement these principles at the fundamental level. The management of the corporate space and the quality enhancement of the products are two crucial segments which demand more transparent and holistic change. This could also lead to a new developmental opportunity where gender-responsive products and services will open up new avenues for new technologies. Application of the various methods of analysing gender principles in AI can also be a new resource for gaining new knowledge and developing effective technologies. Reducing biases is reducing risks which will inevitably lead to better organisational support and institutional mechanism for the corporate world. Proper training of the engineers and continuous training is essential to cope with the changing social scenarios and the rapid technological changes. Integrating the knowledge from the gender experts and social scientists regarding developing and managing AI technology should be a conscious process where the person is adequately aware and trained about the AI functions and understand how AI technology is developed and deployed. The organisation should encourage strict principles in having gender equality and women's rights experts in panels and the continuous development of the software and hardware. Frequent involvement of technical experts in various social situations and workshops is essential to have a first-hand experience of the world that the engineers are trying to replicate. These concerns led to a broader construction of the technological world which ought to be more inclusive and non-discriminatory. It is not only about fixing the bugs within the system but also fixing them from outside: a more socially responsive development of technology. Getting more into the missing links and focusing more on the intersections, the principles and ethical parameters made by the organisations should address the unequal power structures of the technology sector. Lawfully taken positive actions, parities in training, education, retention, and promotion need to be fostered for an inclusive culture within AI organisations.

10.5 Conclusion: Digital Inclusion, a Final Call

The time is for a cultural call towards technological awareness and change. A new narrative of competitiveness needs to be built where more gender-sensitive technology will have more reputational gain rather than technology falling into the product trap of profit only. The seeds of technology are gendered, as can be seen currently. In a posthumanist imagination of the world, technology cannot be a post-women or post-gender. Whereas biological consideration of sex will be irrelevant for a machine or system, choosing gender and other related considerations will be a hermeneutic call. The identities of the robots or systems will be defined by a more contextualised process based on the human-computer interactional model. Cultural beliefs and economic, social, and political interests will play significant roles in defining the future of technology and its implementation. As human conditions constantly evolve, the epistemologies of the technological fields also need continuous scrutiny to comply with the new symbolic orders. While having more female representatives in AI technology is an effective solution, more is needed to change the prevailing scenario for so long. Retaining the female in principle role is a long-drawn affair that needs a severe social and political overhaul. Referring to the male dominance in the AI industry of *Silicon Valley*, Emily Chang titled her book, *Brotopia: Breaking Up the Boys' Club of Silicon Valley (2018)*, discusses how the industry is already laden with so many male engineers and practitioners [4]. It is not an issue exclusively of dominance or discrimination, but as argued earlier, it concerns the representation and fairness of the workplace. While discussing gender equality in AI technology, it starts with breaking away from this power structure where 'brotopia' is nullified with an equal distribution of work and significant positions in the industry between two genders.

AI industry and machine learning are long-evolving bodies of work, and in the coming decades, they will gradually take a more concrete shape. The problems with gender are not any more the older problem of discrimination at the societal level or earlier forms of violence. The computer and, especially, AI have brought the issue to a more concrete level, embracing multiple phenomenological sets of practices operating at different levels of technology applications. The time calls for a continuous focus on the various aspects of their development, not only the gender aspects but also race, religion, and other social aspects, to have a transformative technology which contributes to our being and becoming. As the whole world of technology is gradually moving from 'humans to algorithms' [9], we also need to have an equal and opposite direction where the big data algorithm cannot undo the nuances of human liberty and freedom. The drama of decision-making in life should be a human affair where technology becomes the enabler for a better and quick decision. In the face of liberal democracy and global market capitalism, it is even more important that human beings remain aware of the digital dictatorship. The globalisation of technology has benefitted society in numerous ways, but it also obliquely refers to several growing inequalities operating between societies. The concerns around gender in AI technology are one of the crucial components of discrimination that the developmental pursuit and global market growth can leave behind the curtain. The association of the next generation in the techno-field and emerging entrepreneurship will shape the time ahead. Therefore, a pedagogical pursuit of giving teens early exposure to oncoming technology and different nuances of social goods can create a better world for tomorrow. Between the despair of a possible world of post-human societies and a hope for a technologically sound society, we still have a choice. Whereas the twentieth century revolved around reducing the social inequalities of gender, class, and race, this new era of machine learning demands a similar effort rooted in algorithms and computer interfaces. Before the big data giant catches up and manipulates the path to human cognition and conscience, before our choices fall into digital algorithms completely, before the deep secrets of our life become a knowledge of the computer brain, a holistic effort is necessary to shape our technological development for the future. The technology that disrupts the world can equally become an engineer to life-like problems and challenges. As social tensions intensify world-over and global wars are looking for more secret weapons, technology needs to be placed securely and safely in a humane world of ethical development.

The future is not bleak; rather, it is a visionary time of possibilities. Nevertheless, there is a particular urgency to these concerns around gender violence, which need to get into the action plan and framework of the technological enhancement ahead. The landscape of future technology is enormous. To build an ecosystem that equally contributes to everyone, we need an ethical boundary that not only restricts but shows the necessary diversions and moves that are required. It is a global concern which needs a global intercultural dialogue beyond all boundaries of national and regional communities. It is a call for observation, investigation, and choice. Whereas on the one hand, it threatens to undermine the social order and disrupt harmony, on the other hand, it also shows its enormous potential at shifting the nature of authority and bringing a new sense of technoliberalism: a decentralised technology with minimal control on ideas and practices of liberty, individuality, responsibility, and self-awareness.

References

1. Abril, D. (2019, December 12). A.I. might be the reason you didn't get the job. *Fortune*. https://fortune.com/2019/12/11/mpw-nextgen-ai-hr-hiring-retention/
2. Butler, J. (1990). *Gender trouble and the subversion of identity*. Routledge.
3. Bloomberg. (2022, January 26). Bloomberg's 2022 gender-equality index shows companies increasingly committed to reporting esg data. *Bloomberg*. https://www.bloomberg.com/company/press/bloomberg-2022-gei/
4. Chang, E. (2018). *Brotopia-breaking up the boys' club of silicon valley*. Portfolio.
5. Cowls, J. (2021). AI for social good': Whose good and who's good? Introduction to the special issue on artificial intelligence for social good. *Philosophy and Technology, 34*, 1–5. https://doi.org/10.1007/s13347-021-00466-3
6. Cowls, J., Tsamados, A., Taddeo, M., & Floridi, L. (2021). A definition benchmark and database of AI for social good initiatives. *Nature Machine Intelligence, 3*(2), 111–115. https://doi.org/10.1038/s42256-021-00296-0
7. Floridi, L., Cowls, J., King, T. C., et al. (2020). How to design AI for social good: Seven essential factors. *Science and Engineering Ethics, 26*, 1771–1796. https://doi.org/10.1007/s11948-020-00213-5
8. Gurumurthy, A., Vasudevan, A., & Chami, N. (2019). *Born digital, born free? A socio-legal study on young women's experiences of online violence in South India*. IT for Change. https://itforchange.net/sites/default/files/1662/Born-Digital_Born-Free_SynthesisReport.pdf
9. Harari, Y. N. (2018). *21 lessons for the 21st century*. Random House.
10. Hay, Z. (2019, May 13). Towards trans-inclusive AI: The trouble with gender binary algorithms. *Towards Data Science*. https://towardsdatascience.com/towards-trans-inclusive-ai-a4abe9ad4e62
11. Kaplan, A. (2022). *Artificial intelligence, business and civilisation: Our fate made in machines* (1st ed.). Routledge.
12. Kaye, J., Whitley, E. A., Lund, D., Morrison, M., Teare, H., & Melham, K. (2015). Dynamic consent – A patient interface for twenty first-century research network. *European Journal of Human Genetics, 23*(2), 141–146. https://doi.org/10.1038/ejhg.2014.71
13. Krentz, M., & Yousif, N., et al. (2021, March 17). The payoff for upskilling women in STEM. *BGG*. https://www.bcg.com/en-in/publications/2021/impact-of-skill-building-opportunities-women-in-stem

14. Litosseliti, L., & Sunderland, J. (2002). *Gender identity and discourse analysis*. John Benjamins Publishing.
15. Madgavkar, A. (2021, April 7). *A Conversation on artificial intelligence and gender bias*. McKinsey and Company. https://www.mckinsey.com/featured-insights/asia-pacific/a-conversation-on-artificial-intelligence-and-gender-bias
16. McFarlane, D. (1999). *Interruption of people in human-computer interaction: A general unifying definition of human interruption and taxonomy*. Storming Media.
17. Millett, K. (2016). *Sexual politics*. Columbia University Press.
18. Mohanty, S., & Bhatia, R. (2017, August 24). Indian court's privacy ruling is blow to government. *Yahoo News*.
19. Mollin, S. (2012). Revisiting binomial order in English: Ordering constraints and reversibility. *English Language & Linguistics, 16*, 81–103.
20. Nissenbaum, H. (2009). *Privacy in context-technology, policy, and the integrity of social life*. Stanford University Press.
21. The United Nations Educational, Scientific and Cultural Organization. (2020). *Artificial intelligence and gender equality-key findings of UNESCO's global dialogue*. UNESCO. https://unesdoc.unesco.org/ark:/48223/pf0000374174
22. The United Nations Educational, Scientific and Cultural Organization. (2019). *I'd blush if I could: closing gender divides in digital skills through education*. https://unesdoc.unesco.org/ark:/48223/pf0000367416
23. Vinuesa, R., Azizpour, H., Leite, I., et al. (2020). The role of artificial intelligence in achieving the sustainable development goals. *Nature Communications, 11*, 233–234. https://doi.org/10.1038/s41467-019-14108-y
24. Wachter, S., Mittelstadt, B., & Floridi, L. (2017). Why a right to explanation of automated decision-making does not exist in the general data protection regulation. *International Data Privacy Law, 7*, 76–99.
25. Wang, Y., & Kosinski, M. (2018). Deep neural networks are more accurate than humans at detecting sexual orientation from facial images. *Journal of Personality and Social Psychology, 114*(2), 246–257. https://doi.org/10.1037/pspa0000098
26. West, J. (2014). *Cyber-violence against women*. Battered Women's Support Services. www.bwss.org/wp-content/uploads/2014/05/CyberVAWReportJessicaWest.pdf
27. World Economic Forum. (2018). *Global gender gap report 2018: Assessing gender gaps in artificial intelligence*. https://reports.weforum.org/global-gender-gap-report-2018/assessing-gender-gaps-in-artificial-intelligence/. Accessed 6 June 2022.

Chapter 11
English Lecturers' Digital Resources Use at Universities of Nepal Amidst Unsecured Online Environment

Eak Prasad Duwadi, Siddhant Koirala, Dipin Ale Magar, Susan Shrestha, Saman Adhikari, and Ashmita Chapagain

11.1 Introduction

Teachers are no longer expected to stand in front of students with a book in one hand and chalk in the other. Approaches to teaching and gaining knowledge have altered dramatically because of societal changes and the speedy boom of digital technology. Over the final few decades, the position of motivation and creative thinking has ended up being extra recognized. A trainer has a responsibility to help students boost motivation and creativity. There are many distinct ways and textbooks to choose from. Information and verbal exchange technology (ICT) are additionally important tools that are an advantageous way to help scholars end up more motivated.

Crystal [8] examines the exceptional changes that have introduced computers, cell phones, and other electronic devices into the world of private digital assistants and cell phone answering services in our everyday school room in addition to the applied sciences that have been used by instructors since the mid-twentieth century. If a teacher wants to excite and encourage students, they have to maintain up with technological advancements. They inspire you, they delight you, and you wind up studying a ton even when you are not be aware of it.

Digital natives are accustomed to obtaining facts quickly. They multitask, select games to tough labor, and thrive on speedy pleasure and ordinary rewards [16, p. 2]. Despite the captivating sources and techniques, however, the hole between bright youngsters and youngsters with learning difficulties has been consistently widening.

In Nepal, we are confronted with a shortage of fabulous materials to help college students with disabilities in studying an overseas language. Sukying [21] defines language variations as "students with language studying issues" and provides a clarification for the distinction between language variations and language learning

E. P. Duwadi (✉) · S. Koirala · D. A. Magar · S. Shrestha · S. Adhikari · A. Chapagain
Kathmandu University, Kathmandu, Nepal

D. Mishra et al. (eds.), *Communication Technology and Gender Violence*, Signals and Communication Technology, https://doi.org/10.1007/978-3-031-45237-6_11

impairments. He claims that children with language variations may now not do as well as their friends, because they lack cultural and linguistic experiences, as well as a limited vocabulary due to confined publicity to hearing and using English. These pupils switch from one language to every other inside a single sentence when communicating.

Their nonverbal competencies (gestures, facial expressions, and physical proximity) are, on the other hand, age appropriate. Despite the fact that Sukying [21] is referring to students who study English in an English-speaking surroundings but whose native language is now not English, the description of such students can be utilized to our situation.

Many students in Nepal, for example, who are part of groups of students with learning issues, fit the description. "Educators have difficulty distinguishing language differences from disability when explaining the academic struggles these students face," according to Sukying "school officials report a lack of tools, procedures, or qualified staff to adequately identify these students and their needs." The author, like before, speaks to the situation in the United States. However, similar issues may arise in European or Asian school settings.

While information technology is one of the greatest ways to create and increase motivation in the students, the current thesis focuses on the use of ICT tools as an assistance in teaching English as a foreign language to the students. The research is about the use of information communication and digital resources among the teachers of English in Nepali Universities because there has not been research about this topic. The use of ICT in the classroom will result in a wide range of English material, contexts, and pedagogical techniques. ICT enhances the dynamic, adaptive, and innovative nature of the English language environment [5]. However, this study focuses on the use of the ICT tools by teachers of Nepali University while teaching their students. Therefore, it is seen that this research is different from the ones done previously elsewhere.

The proportion of these with liberal arts tiers is declining as the majority of them at an increasing number choose career-oriented Majors such as business and Engineering. New teaching and learning are still at the heart of any discussion of academic accomplishment. ICTs give educational institutions and other organizations a chance to harness and use technology to supplement and support the teaching and learning process. Despite widespread support for ICT-assisted teaching and learning, as well as investment and donations of ICT equipment to Nepali University, the university still faces the challenge of transforming students' learning processes to equip students with the skills they need to function effectively in this dynamic, information-rich, and ever-changing environment.

The cause for concern is that until this issue is solved, the University's investment in ICT development may be wasted, and progress in improving teaching and learning quality can be slow. This may lead to the University failing to meet its mission and producing graduates who are prepared for a world of work that is increasingly reliant on ICT-assisted knowledge development and distribution. In light of this disparity, it is necessary to investigate the specific effects of ICT resource availability, threats, and usability in Nepali University students.

11.2 Methodology

A research sketch is a quintessential approach that directs the lookup project's facts gathering and analysis phase. It presents a framework for defining the sort of statistics to be gathered, as nicely as its supply and collecting processes [7]. This study is qualitative. It is designed to find out a teacher's relation to ICT use when educating normal students with qualitative protocol, which were designed like interview questions.

With a set of questionnaires, the respondents had been interviewed to discover the teacher's relation to ICT use when instructing regular students. In order to answer the questions, online questionnaires were created for teachers. The selection of participation used to be unhurried. The author's non-public contacts as well as the online lists of Universities in Nepal have been used to locate participants for the study. Participants for the study had been recruited using the non-public contacts as properly as online lists of Nepalese Universities. The data was once gathered through the use of free online surveys and due to the pandemic face to face interview ought to now not be possible. In addition, as the net was once open to Nepal, teachers should have accessed the equal materials, net sites, and directions online. There were 30 complete teachers who were from exceptional Nepali Universities.

The find out was once carried out in a moral manner, with no biases. This learns about adhering to the key standards that must be addressed when doing research. This finds out about did not jeopardize respondent confidentiality. Prior to finishing the study, the respondents would be asked to complete a permission form. The researchers ensured that the inquiry would now not discriminate against persons or respondents primarily based on their caste, religion, age, financial level, or preceding lifestyle experiences. The findings of this study would be introduced besides any records manipulation. Finally, if any one was unwilling to participate in the research, the researchers would no longer compel them to be the respondents.

11.3 Findings

Out of 30 respondents, 90.5% had been male and 9.5% had been female. Similarly, 14.3% of them were of the age crew 26–35 years, 47.6% had been of 36–45 years, 33% had been of 45–55 years and 4.8% had been of the 56–65 years age group. Among all the instructors approached 90.5% were the usage of ICT and 9.5 percent had been not the use of any ICT while educating in the class. When asked about if they use ICT and if not why, then 12.5% of them replied that they were not using ICT because of lack of time and all of the rest have been using it.

The teachers answered that 80% of them had been the usage of laptop and computers as a medium of ICT, 15% had been using projection technological knowledge and the last 5% used the cell phones. Out of the surveyed university professors, 75% each used the ready-made and self-made educating materials, 20% of them never used the ready-made material, and solely 5% of them always used the ready-made materials. While teaching through ICT in class, many troubles manifest occasionally. And for the duration of those moments, the teachers want to adopt several techniques to tackle them.

When the English lecturers had been asked about their adoption strategies for ready-made venture and materials for the students, 38.9% of them said that they decreased the variety of ideas, 16% of them used choices such as position taking part in and dramatizing, 16.7% stated that they used chart or table, 11.1% stated that they highlighted the materials, 11.1% spoke back that they confirmed the equal content in another medium like movie, filmstrip, tape, and many others and 5.5% of them in no way used any adoption strategy.

The teachers also faced unique sorts of troubles whereas utilization of ICT was associated with technical features. When requested they have a variety of problems. 30% of the total instructors faced technical equipment unreliability, and 25% of the complete instructors confronted some portions of tools that could be tough to handle. 10% of the complete teachers depend on ICT tools a lot and again 10% of the whole instructors confronted a whole lot of pre-work, 15% of the whole instructors encountered college students taking part in video games or striking round while doing a task on the internet. Five percent of complete instructors encountered their college students going through generic power cuts and bad connectivity whilst some other 5% of whole instructors encountered their students' dealing with issues in uploading assignments on time, delayed submission due to lack of acceptable web service or due to use of negative devices.

11.4 Discussions

The enhancement in the use of technology has taken a fast pace that we can't even imagine our day besides it. It has made our existence comfortable, yet state-of-the-art and complex in many aspects. Along with the advantages of using technology in our everyday life, it has also delivered some serious challenges in the shape of cybercrime, cyberattacks which have become a primary problem in the modern-day world. As many people now have got entry to the cyber world and social networking sites, cyber violence toward gender is on the rise.

Internet customers do not feel protected online. They ride threats associated with identity theft, malware or viruses, protection of monetary information, and phishing attacks that may harm their professional recognition and private lives. Online chance perceptions stem from personal experiences, others' experiences, and the news media, as a consequence, lead to a practical grasp of online security threats. Although Internet customers use built-in gadget settings (e.g., antimalware software, firewall, and computerized updates) to keep their online protections, three in five humans do not believe they can be absolutely nameless online and they are an increasing number of involved about their private facts online [14].

Ninety-nine percent of computer systems are susceptible to threat attacks as a result of the incidence of Adobe Reader, Adobe Flash, and Oracle Java [25]. Despite high risks associated with frequent computer and technology activities, Internet customers take very few moves to defend their computers, smartphones, and tablets. Moreover, they interact in things that jeopardize their online security and reputation, such as posting records that ought to be misused with the aid of online predators [14].

Threats to online security translate into cybercrime, with U.S. economic losses ranging from $24 to $120 billion yearly [23], making online security a policy and academic priority. Taking into account the discrepancy between realizing threats and taking shielding actions, Today's learn is about applications Theory of Protection Motivation (PMT) to apprehend what drives online safety behaviors in the context of home pc use. It contributes to research on computer systems and protection by way of integrating an often not noted PMT variable: prior experience.

11.4.1 Cyber Issue and Security

Online violence is emerging as a world problem that widely impacts women. To cite Britanica [3], violence is an act of physical force that causes or is supposed to motivate harm. The injury inflicted by using violence may be physical, psychological, or both. Violence may also be exclusive from aggression, a greater normal type of hostile behavior that may be physical, verbal, or passive in nature.

There is enough evidence to exhibit that ladies web customers are a problem to time-honored violence, harassment, hate speech, and censorship. Cyber abuse concentrated on ladies is a gender-based violence (GBV) perpetrated by way of the use of technology. They have been the victim of technology-based abuse such as intimate accomplice violence, stalking, harassment, and picture-based sexual abuse. Due to the lack of attention to gender in cyber security, it has created an environment where misuse of science is common or at least tolerated.

Although cyber security and gender violence is regularly the issue of public debate, rising tech like artificial genius is not only with the gendered impacts, other more without difficulty available and huge tech such as social medias, SMS messaging, and features designed to come across the misplaced gadgets are extra many times use to abuse tech in the shape of gender violence. Technology-facilitated abuse or "tech abuse" through global positioning system (GPS) trackers, smartphone apps, or structures such as Facebook has a massive impact on the nature of intimate associate violence (Slupska & Tanczer, 20, p. 654). Another instance is the global positioning system (GPS) in telephones and cars that are used to stalk, harass, and abuse females.

According to Aneesh [2], "Women are more probable than men to be the victims of extreme types of cyber violence and the influence in their lives is a long way more traumatic" (EIGE, 10). Women are disproportionately vulnerable to special forms of online abuse in a variety of components of the world, especially girls of particular religious, ethnic groups, race, sexual orientation, economic reputation, and with disabilities.

Paste your text here and click on "Next" to watch this article rewriter do its thing. `A ballot with the aid of Charman [6] revealed that about one-fourth of the 4000 girls surveyed in the United States, United Kingdom, Denmark, Sweden, Spain, Italy, and Poland experienced some shape of online abuse at least once. What is more, 41% of these women who experienced online abuse feared for their non-public security because of their abuse and harassment (Amnesty International, 2017).

Women have gotten frightening messages, risks of savagery, and physically specific immediate messages, messages, pictures, and recordings by using the ability of dating, virtual entertainment, and different web-based stages, as well as in discussion channels and texting administration. The analysis offers a stepping stone for in addition cyber security-centric critiques and tackles the possible misuse of applied sciences from "within." Having outlined a clear set of shortcomings in the existing responses to the trouble of tech abuse, our dedicated threat mannequin may also guide future science design.

Women have been threatened with sexual violence, as well as sexist, misogynistic, discriminating, and biased statements, all of which have been sent to them with the aid of ICT, developing hostile surroundings for them online. Many experts have warned now not to take cyber violence as an awesome problem from actual world forms of gender violence. Females whether women or girls, who have been sexually harassed, abused, or violently assaulted through an intimate partner offline are additionally victim of cyber abuse perpetrate via the equal individual.

In conjunction with the accelerated adoption of technology, there is an increase in its misuse, particularly in the context of home and sexual violence [20]. Women's lives are extensively impacted by cyber abuse, which takes several forms. Perpetrators are well aware of their victims; the information is accumulated through a number of websites and portals. However, the victims hardly ever recognize the other side, privacy insurance policies, and safety measures [20].

11.4.2 Threats and Opportunities of Online Classes

The Educational way of life in Nepal has been dominated by face-to-face tutoring that has a lengthy history beginning from the Gurukul way of life to the existing formal schooling. In the Gurukul system, college students used to go to a Guru's Ashram (teacher's domestic or temple) where the Guru used to train religious and nonsecular knowledge and lifestyle skills. However, emerging practices of the usage of technology in schooling have been promoting Online Learning (OL) as a structure of distance training and, of late, it has been gaining recognition among Nepali students [15].

After the COVID-19 pandemic, the nations such as growing countries like Nepal are slowly getting into the online environment. But the question usually stays about the protection of ladies in the online environment. However, we can see loads of benefits of those online presence for each gender but the threats additionally exist in a massive number. As a woman, she gets the opportunity to utilize her best time to learn new skills, get know-how about the specific matters, and be up to date about the world's current affairs. But can we make certain that their presence is protected and secured? Probably no longer due to the fact of the stay conditions that we see currently in Nepal especially.

Females, what we call the second vital part of the society after male includes women and girls too. Women and ladies are usually dominated with the aid of male groups and they have the sort of understanding concerning girls that the women's team are isolated and made to experience on their own and pressured too. Online sexting, bullying, blackmailing for bodily relationships, and some of the few but tremendously practiced online crimes and not to doubt girls are the victims of those. And due to the fact of these as nicely we can say, the online structures are any other region for female violence and making them mentally weak. In the context of Nepal, the technology area is dominated by the way of male and female individuals in such technical fields are considered from a one-of-a-kind however negative perspective. So on choosing professions there are restrictions up to positive extinction.

11.4.3 Security Threat and Protection in Online Gaining Knowledge of Ecology

Nepal is no longer so ahead and advanced with regard to net get right of entry to and devices on hands. Accessibility of the net is higher in Nepal however training regarding cyber protection and environment-friendly use of social media barring developing disturbance in the digital world is lots less in practice. Thinking dimension is much narrower and content material in social media is a whole lot disturbing, from time to time it even spreads real-life violence. People accept as true with the faux information and also share this information as it is tough to differentiate fake or true. Lack of schooling and recognition of the smart use of social media are hindering the peace of a society due to which a number of inhuman activities are also rising up. These issues need to be addressed and need to be concerned to resolve earlier than it takes a shape of battle to supply a secure platform for ladies with the help of a variety of authorities.

Teacher schooling packages in Nepal have frequently been criticized for no longer sufficiently addressing the wishes of the numerous students which hosts extra than a 120 linguistic and cultural groups. Oftentimes, it has been pointed out that trainer training curricula borrow ready-made pedagogical models from the developed world except thinking about their appropriateness in Nepali classroom contexts. This act of borrowing is thought to be disempowering for the Nepali teacher training neighborhood on two grounds: first, Nepali civilizations are poor of their very own pedagogical models; and second, neighborhood lacks critical knowledge and capabilities to tackle the issues of exclusion, inequity and meaninglessness in instructor education. Likewise, the thoughts of online, flexible, distance, and ubiquitous mastering are regularly credited to be the beginning of the developed world.

One type of cyber violence is cyber stalking. An individual sends the victim undesired emails and text messages on a normal basis. Another type is cyber harassment, which involves making threats of violence, hate speech, or sexually explicit communications to someone by the internet. Victims are additionally threatened with their pretend specific pix and pornographic videos. Posting sexually specific snapshots or movies online barring the consent of the man or woman is regarded as non-consensual pornography. Non-consensual pornography, which is usually done by way of the former partner, is one of the most horrific types of cyber abuse that many girls suffer. According to studies, about 90% of non-consensual pornography victims are women, and the incidence rate is rising.

The influence of such cybercrimes on women is frequently main to the suicide through many women. They are being blackmailed and abused with the aid of their personal companion multiple times. Due to this, they have come to be effortless victims of cybercrime. This frequently leads ladies to the state of stress, nervousness, and depression. Furthermore, greater than a 1/3 of folks who were subjected to cyber harassment stated they felt anxious, and a fifth stated they referred to modifications in their dozing and ingesting routines, as nicely as a sense of helplessness [22]. They have been secretly abused and dominated by the participants of their family or relatives. Their rights have been violated and they have been suppressed via males many times. They are afraid to communicate up and elevate their voice for justice. As a result, they feel insecure in the society and they have been suppressed and victimized via many human beings and their very own relatives.

11.4.4 COVID-19 Impacts

The COVID-19 pandemic has extensively accelerated the risks of gender-based abuse on the web and in the actual world. Leisure time was once spent scrolling social media by using most of the humans and human beings with horrific intentions used the skill for bad purposes. Facebook continues to be the most frequent channel for gender violence at some point of the world, observed by Instagram and WhatsApp.

Multiple sorts of cyber harassment, such as trolling (continuous use of abusive terms or language) account hacking, and being emailed sexual pictures besides authorization, have constantly been reported on such sites, with the majority of victims being women. The victims received threats generally from strangers who commonly demanded specific photographs and videos.

In the past few years online gaming has ended up extra liked by means of teens and youths which in addition received more popularity throughout COVID-19 pandemic. Also the modernized world has taken over the bodily gaming periods and human beings often decide upon online video games over bodily games. Online gaming world consists of activities that encompass reel violence in the course of gaming classes so male are considered suitable and greatest over female. Female gamers were not handled respectfully and also received harsher comments.

Also the game programmers make most of the male characters in the recreation and are given supreme strength over females which is the important motive developing teens to assume the online game is solely viable for boys. The internet of records technology and software developers consist of most adult males and females are always behind them. Due to much less girl tech handlers and developers male dominant in this field and ladies are regarded much less sizable along with discrimination.

Body perfectionism is something poisonous scattered by social online websites that hit hard on women. Women are regularly imagined to have a perfect physique like some of so-called social media influencers. Influencers additionally manipulate their followers with a variety of products and their followers have blind imitation. An estimated 45 million people follow diets every year and spend $33 billion on weight-loss items, according to Boston Medical Center. These influencers have a global impact, and because of the pandemic, fewer individuals are working, which reduces their previously greatest effect.

To mitigate the direct influence of virus and stop the students from becoming psychologically depressed, academic establishments have initiated the momentum of teaching the students through different skills and modalities based totally on each human and fabric resources handy in the context. In this context, the virulent disorder has made college students to be at home, which has led numerous institutions to arrange a variety of e-learning degrees (even though with confronts) to help and make positive that the tutorial schedule runs to closing stages. Technology-based totally educating mainly online schooling has ended the most appropriate choice to keep academic activities functional in many parts of the world at some stage in the pandemic period [15].

This digital world has made women and men assume twice and every now and then even extra earlier than leaving domestic, creating intellectual disturbance, self-doubt, low self-esteem, and fear of being judged. People make physical facets as judgmental parameters. According to the Girls' Attitudes Survey conducted in 2016, 47% of females aged 11–21 disclosed that the way they appear holds them back and limits what they can do. Unfortunately, physique shaming has permeated most people's everyday lives and experiences. The identical survey claims that 94% of teenage females have reported being body shamed and 57% of 12–20-year olds assume that the cause of their bullying used to be due to their appearance.

Gender violence in digital structures takes place in three ways. In the beginning, social networking websites may additionally cause more abuse. To harass the victim, the same person establishes many accounts. Second, the graph of digital apps fosters cyber aggression. Abusers, for example, can also use Snapchat's map functions to display their victims in super detail. Third, the anonymity afforded by means of social networking websites has enabled people to make derogatory remarks and harass the man or woman (Abdullah, 2021). According to recent studies, ladies with prominent social media profiles are disproportionately targeted. This indicates that the greater a well-known woman's account is, the extra prone she is to be a victim of online harassment. When this involvement becomes abusive, this may imply that the structure of these structures is itself a facilitator of gender violence.

To quote Paudel [15] again, nonetheless, it is believed that the current study has introduced a wave of lookup in online schooling and the associated phenomena in the context of Nepal. Moreover, it presents feedback to the instructors and college students to undergo applicable characteristics for online training and for the policy-makers, curriculum designers, and textbook writers to strengthen gorgeous ICT policy, and ICT-friendly curricula and guides suitable in the context of Nepal even after the fear of pandemic COVID-19.

11.4.5 Online Security as a Challenge

Security has been the most challenging due to COVID-19 as many youths were involved in unethical things like hacking, stalking, and developing fake accounts. Various social media money owed from others had been hacked and posting of false information about the usage of those potential was also found for the duration of the time of pandemic but the essential target of this trouble was once female. In Nepal girls have a tendency to have lower hierarchy than male participants of society so there usually exists the query mark in front of women's reputation. Targeting these accounts and blackmailing girls used to be much simpler for those digital frauds so this form of undertaking took incredible heights in the course of the lockdown period.

Despite such heinous acts of harassment and abuse directed at women, it is imperative to highlight how female and different victims can also utilize digital systems to raise their voices and recover company over the issues. But the price is very low. While the privacy of Facebook, Instagram, Twitter, and WhatsApp has its pitfalls, it additionally offers women a safer digital and bodily organizing platform. Furthermore, people might also utilize such forums to raise their voices and elevate awareness towards abuse except worry of societal or nation retaliation.

For example, Instagram account has supplied a safe area for survivors to discover their community via "Assault Police" which archives and shares incident narratives of intimate accomplice violence and sexual harassment in Egypt. In a similar manner, Twitter has sponsored social moves aimed at elevating recognition of women's rights and online abuse against women. The hassle will now not be solved clearly until social media corporations assume a larger duty for preventing harassment, threats, intimidation, violence, and abuse.

This study relies on primary sources of records gathered by an online poll. A set of questionnaires was created, disseminated, and statistics used to be gathered through the assistance of the net and these responses are analyzed diagrammatically and explained. Demographic data suggests that most of the teachers in Nepali Universities are male which makes up about 90% of the total instructors and least teachers in Nepali Universities are girl which makes up about 9 point 5 percentage of the total instructors and the age crew from 36 to 45 makes up to 47 factor 6 percentage of the composition, age crew from 45 to 55 makes up to 33 factor three of the whole teachers, age crew structure 26 to 35 makes up to 14 factor three of the complete instructors and age team from 56 to 65 makes up to four point 8% of the total teachers.

Teachers who use ICT make up 95 of the total teachers. ICT when instructing students with language differences makes 9 factors of the complete teachers. Therefore, most of the teachers use ICT but the least number of teachers additionally use ICT when educating college students with language differences. Teachers using ICT make 90 point 5 of the whole teachers. ICT when teaching college students with language variations makes 9 factor 5 of the complete teachers, consequently most of the instructors use ICT but the least wide variety of teachers also use ICT when teaching college students differ.

Thirty-eight point nine of the whole instructors decrease the quantity of ideas, 16 factor seven of the whole teachers furnish alternative methods for students such as function enjoying and dramatizing, 16 point seven of the whole instructors grant a chart graph or table, 11 factor 1% of the whole instructors provide highlighted material, 11 factor 1% of the complete teachers appears for the same content material in another medium (movie/filmstrip/tape) and 5 factor 5% of the whole teachers never used any adaptation-strategies. Thirty percent of the whole instructors confronted technical equipment unreliability; 25% of the complete instructors faced some pieces of tools can be difficult to handle.

Ten percentage of the whole instructors count on ICT tools to a good deal and again 10% of the whole instructors faced too an awful lot pre work, 15% of the total teachers encountered college students taking part in video games or hanging round while doing a venture on the internet, 5 percentage out of total teachers encountered their scholar facing typical energy cut and poor connectivity whilst different 5 percentage out of total teachers encountered their students going through issues in uploading assignments on the time, delayed submission due to lack of suited web provider or due to use of bad device.

This section investigates whether or not or no longer the outcomes of the literature opinions and the lookup findings had been consistent. The results of the survey of Wikan and Molster [24], as nicely as Brodin and Lindstrand [4], 86 disclose that teachers also experience a lack of ICT confidence, frequently no matter having taken part in ICT publications [24, p. 209]. Brodin and Lindstrand [4, p. 86] go further, declaring that educators are mindful of their personal need for training, simultaneously being pissed off with the aid of the lack of time, economic resources, and inadequate response of the management.

The most essential discovering that Wikan and Molster [24, p. 209] concluded from their survey used to be "to integrate ICT in one's very own educating is a hard and gradual process, and instructors should be given time to discover their very own way to merge ICT with their personal instructing style." The consequences of the study carried out by Sanchez and Aleman [19, p. 914] agree with the findings of Wikan and Molster [24], noting that teachers keep an open mind about integrating ICT into their day-by-day practices.

According to the outcomes of the survey via Donnelly et al. [9, p. 1477], instructors who see ICT as a possibility for them to do something new and fascinating with their college students in phrases of how their students learn, push to have a larger range of assets in their lecture room and it includes ICT. Hsu [12, p. 847] examined the relationship between teacher and scholar utilization of ICT and found out that instructor use of ICT extensively determines how often they assign college students ICT activities. It turns out that if a trainer frequently uses ICT tools, they frequently assign college students the same type of ICT activities.

For example, teachers who create difficult multimedia substances are likely to assign college students multimedia activities. It shows that the teacher's very own ICT practices have an effect on the type of ICT activities they assign to students. Samuelsson [18, pp. 15, 15] carried out a study on the use of ICT among 13-year-old Swedish children, "who can be categorised as belonging to the team referred to as the digital generation." The outcomes disclose that all contributors have got the right of entry to ICT however they use it in more than a few ways.

Samuelsson [18, p. 15] found out that the participants' use of ICT differs in both qualitative and quantitative ways, however there is a lack in fundamental laptop skills and seriousness about ICT use as a tool for learning. The findings reveal that school-related computer things to do are alternatively exclusive among the members of the survey and teens prefer listening to music, socializing with pals or playing video games alternatively than the usage of ICT for educational purposes. This additionally stresses the importance of a teacher, who would guide college students towards different ICT preferences that college students can use in or out of school.

On the different hand, according to the find out about by Hennessy, Ruthven, and Brindley [11], students realize the benefits of computer-based equipment and assets as assist and companion the use of such equipment and sources with modifications in working ambience and school room relations, as properly as with raised interest and accelerated motivation inside themselves. It shows that while outdoor students manipulate technology, school-based learning is primarily guided by teachers, timetables, and curriculum requirements.

Abbot and Withgott [1] helped college students and instructors see how ICT tools can contribute to socializing, whilst getting to know something new. The survey yielded the identical result. ICT tools are really helpful while educating students, teachers may also encounter technical troubles while the usage of it and the use of ICT equipment by college students have to be well-supervised by using teachers.

11.5 Conclusion

The pretty existing survey exhibits that materials really are often adapted using adaptation strategies in a delicate way by using offering alternative techniques for college students such as function playing and dramatizing, imparting a graph chart or a table and highlighted substances as well and very international locations of Europe additionally use web quests whilst Nepali instructors for all intents and purposes opt for making find out about tips and worksheets with the assist of ICT tools such as computer systems or laptops in a commonly most important way.

Many instructors additionally particularly agree that the use of ICT-related duties and activities in instructions requires too tons pre-work and the outcomes generally are no longer continually what teachers actually assume due to the fact when working on a computer, for example, college students often honestly decide on enjoying online games or chatting with pals rather of dealing with their tasks in an especially principal way. ICT tools are beneficial for teaching students; nevertheless, instructors may also trip technical difficulties when utilizing them, and students' utilization of ICT equipment ought to be carefully monitored with the aid of teachers.

No matter how serious the issues are, solely a few culprits have been tracked and penalized. The complex gadget of the cyber world wishes superior and nice protection structures to assault similar crimes. People need to be made apprehensive about the cyber pitfalls and cyber security especially, teens and females as they're more prone to it. Culprits in the cases are significantly men and victims, certainly though it varies, are substantially females. Legal fabrics that reply to vicious acts in cyberspace are constantly rushed through in an alarmist, rather than empowering, manner.

Cybercrime is ineluctable with the development of technology. Culprits have set up new methods to habits felonious conditioning thru technologies. Crimes that were restrained to certain areas have now gained instigation to unfold throughout the globe via web and techs. The failure of cyber protection to maintain up with the technological revolution has expanded cybercrimes like gender violence on women. The platform which holds the electricity to spread the mindfulness against similar crimes has been misused for abusing and draining. The goods are lifestyle threatening certainly leading to instances of self-murder with the aid of the victims.

11.6 Limitations of Study

Though Nepal has at least 12 universities across Nepal and lots of them are concerned in teaching English in special Universities, it is no longer viable to reach all of them. However, meeting and interviewing 30 English teachers is something that displays credibility as these instructors characterize the complete population. But due to lockdown, the total records were extracted with the help of online forms.

Acknowledgement The lookup was executed with the assistance and aid of a lot of people the authors do acknowledge them all. Thanks to the UGC Nepal Sanothimi Bhaktapur Nepal for providing an SDI project, and to Kathmandu University School of Management for this impressive opportunity.

References

1. Abbot, P., & Withgott, J. H. (2004). Phylogenetic and molecular evidence for allochronic speciation in gall-forming aphids (pemphigus). *Evolution, 58*(3), 539–553.
2. Aneesh, A. (2021). *Virtual misogyny during lockdown: Selective case studies on victims of virtual violence in Kerala* (Master's thesis, University of Kerala). University of Kerala.

3. Britanica. (2022, July 12). *Violence*. https://www.britannica.com/topic/bullying
4. Brodin, J., & Lindstrand, P. (2003). What about ICT in special education? Special educators evaluate information and communication technology as a learning tool. *European Journal of Special Needs Education*. https://doi.org/10.1080/0885625032000042320. Accessed 15 Jan 2014.
5. Cakici, D. (2016). The use of ICT in teaching English as a foreign language. *Participatory Educational Research, 4*(2), 73–77.
6. Charman, T. (2018). Sexual violence or torture?: The framing of sexual violence against men in armed conflict in Amnesty International and Human Rights Watch reports. In *Sexual violence against men in global politics* (pp. 198–210). Routledge.
7. Churchill, G. A., & Iacobucci, D. (2006). *Marketing research: Methodological foundations* (Vol. 199, No. 1). Dryden Press.
8. Crystal, D. (2010). The Cambridge Encyclopedia of Language.Third edition, CUP.
9. Donnelly, D., McGarr, O., & O'Reilly, J. (2011). A framework for teachers' integration of ICT into their classroom practice. *Computers and Education, 57*, 1469–1483.
10. European Institute for Gender Equality. (2017, August 8). https://eige.europa.eu/publicationsresources/publications/european-institute-gender-equality-eige-brief-2017?language_content_entity=en
11. Hennessy, S., Ruthven, K., & Brindley, S. U. E. (2005). Teacher perspectives on integrating ICT into subject teaching: Commitment, constraints, caution, and change. *Journal of Curriculum Studies, 37*(2), 155–192.
12. Hsu, S. (2010). Who assigns the most ICT activities? Examining the relationship between teacher and student usage. *Computers and Education, 56*, 847–855.
13. Javid, M. A., Abdullah, M., Ali, N., & Dias, C. (2021). Structural equation modeling of public transport use with COVID-19 precautions: An extension of the norm activation model. *Transportation Research Interdisciplinary Perspectives, 12*, 100474.
14. Kimble, M. (2016). *Online gendered harassment and violence: Naming the harm and punishing the behavior* (Doctoral dissertation).
15. Paudel, P. (2021). Online education: Benefits, challenges and strategies during and after COVID-19 in higher education. *International Journal on Studies in Education, 3*(2), 70–85.
16. Prensky, M. (2001). Digital natives, digital immigrants part 2: Do they really think differently? *On the Horizon, 9*, 1–6.
17. Rainie, L., Smith, A., Schlozman, K. L., Brady, H., & Verba, S. (2012). Social media and political engagement. *Pew Internet & American Life Project, 19*(1), 2–13.
18. Samuelsson-Brown, G. (2010). *A practical guide for translators*. Multilingual Matters.
19. Sanchez, J. J. C., & Alemán, E. C. (2011). Teachers' opinion survey on the use of ICT tools to support attendance-based teaching. *Computers & Education, 56*(3), 911–915.
20. Slupska, J., & Tanczer, L. M. (2021). Threat modeling intimate partner violence: Tech abuse as a cybersecurity challenge in the internet of things. In *The Emerald international handbook of technology-facilitated violence and abuse*. Emerald Publishing Limited.
21. Sukying, A. (2021). Choices of language learning strategies and English proficiency of EFL university learners. *LEARN Journal: Language Education and Acquisition Research Network, 14*(2), 59–87.
22. Viraja, V. K., & Purandare, P. (2021, July). A qualitative research on the impact and challenges of cybercrimes. *Journal of Physics: Conference Series, 1964*(4), 042004. IOP Publishing.
23. Waterman, A. S. (2013). The humanistic psychology–positive psychology divide: Contrasts in philosophical foundations. *American Psychologist, 68*(3), 124.
24. Wikan, G., & Molster, T. (2011). Norwegian secondary school teachers and ICT. *European Journal of Teacher Education, 34*(2), 209–218.
25. Zaharia, C., Reiner, M., & Schütz, P. (2015). Evidence-based neuro linguistic psychotherapy: A meta-analysis. *Psychiatria Danubina, 27*(4), 355–363.

Chapter 12
Cyberfeminism, Gender Dynamics and Women Empowerment

Navleen Multani

12.1 Introduction

The digitalization of everyday life amplifies the dynamics of politics and gender relations. The smart devices in everyone's pocket increase the possibilities of communication and participation in democracy. The participatory spaces pave the way for feminist development studies. The digital environment provides empowered spaces, invited spaces and claimed spaces to the marginalized. A diverse feminist discourse promotes inclusion of participatory processes in the context of structural inequalities. The discussion on new modes of digital communication in the 1980s is identified as cyberfeminism [2, pp. 232–237]). Cyberfeminist discourses promote "diversity through the digital performance of marginalized identities in digital counterpublic" (p. 238). Cyberfeminism foregrounds the relationship between cyberspace and contemporary feminist movements. Cyberfeminism articulates a range of theories, practices, debates and the relationship between gender and digital culture. There are a number of stances theoretical and political in relation to internet technology which distinguishes between old cyberfeminism and new cyberfeminism. The old cyberfeminism, characterized by a utopian vision of a postcorporeal woman corrupting patriarchy, is different from the new one that confronts top-down from the bottom-up. These variants of cyberfeminism focus on gender and digital technologies as well as the cyberfeminist practices [4, pp. 102–103].

N. Multani (✉)
Jagat Guru Nanak Dev Punjab State Open University, Patiala, India
e-mail: navleen.multani@psou.ac.in

D. Mishra et al. (eds.), *Communication Technology and Gender Violence*, Signals and Communication Technology, https://doi.org/10.1007/978-3-031-45237-6_12

12.1.1 Cyberfeminism

Cyberfeminism, a term coined by the British cultural theorist Sadie Plant, extends beyond academia. It fuses discussion on feminist scholarship, digital art projects and online women's empowerment projects including peer support groups [2, p. 238]. Sadie Plant, Director of Cybernetic Culture and Research Unit, University of Warwick (Britain), in 1994, describes the work of feminists inclined towards theorizing, critiquing and exploiting the internet as cyberfeminism. Plant's *Zeroes and Ones* is a bold manifesto on the relationship between women and machines. As Sadie Plant steers a course beyond the old feminist dichotomies, she explores the emergent entities and shatters the myth that women are victims of technological change. She believes in the potential of internet technologies to transform the lives of women. She envisions a post-patriarchal future and traces the critical contribution of women in the progress of computing from the industrial revolution to the internet times. Plant conceptualizes cyberspace as liberating place for women and symbolically renders 'zeroes' as female and 'ones' as phallic and male predicting the digital future to be feminine, non-linear and distributed in which 'zeroes' are displacing the phallic order of the 'ones' [8, p. 99]. Her optimism about gender equality in cyberspace is a reaction to the previous conceptualizations of technology as masculine. Cyberfeminism opens up space for a dialogue that accommodates technology and gender development. The narrative of progress, equality and democratization is embedded in the concept of development. The consumer culture is a part of progressive enlightened move towards development [7, p. 617]. Rosi Braidotti and Sadie Plant recognize the potential of cyberfeminism as a promising feminist practice in contesting technologically complex territories and chart new course for women [5, p. 1281].

12.1.2 Women and Internet

Economic oppression, lack of access and digital divide deny agency to women on the internet. Despite the fact that women are at the bottom in global economic hierarchy, their internet participation has taken a leap. Gajjala points out that those excluded from the mainstream society want to include themselves in these new technologies on terms of their own. They can see themselves as the protagonists of revolution. New technologies, for women, mean including themselves in the networked global feminism. This global internetworked global feminism bypasses national states, local opposition, mass media, indifference, and major national economic sectors. Thus it opens up a terrain for activism and addresses gender inequality [4, p. 106]. Many women view internet technology as an important medium for movement toward gender equality. Wendy Harcourt, an Australian researcher and author of *Women@Internet:Creating New Cultures in Cyberspace* (1999), writes that internet is a tool for creating communicative space embedded in political reality that empowers women. As internet facilitates transnational feminist networks, a number of associations and organizations have initiated and achieved pursuit of more gender-equitable society. Global feminist networks in South Asia have fostered a challenge to gender-specific abortion or some selection or termination of pregnancy in which the foetus is a female. The mobilization of global awareness and opposition to repressive practices of Taliban by the revolutionary association of women of Afghanistan is also an example of the effective usage of internet by global feminist organizations. There are many women organizations online that have the potential to transform the local conditions of the institutional domains where women are the key actors. Internet is also a safe space for women resisting repressive gender regimes. Fereshteh Nouraie-Simone's notable work, *On Shifting Ground: Muslim Women in the Globalized Era* (2014), describes cyberspace for young educated Iranian women as liberating territory and place to resist traditionally imposed subordinate identity. The structure of interconnections draws participants into the ongoing discourses on issues of gender politics, patriarchy and feminism. Hence, the cyberspace possibilizes agency and empowerment to women [4, pp. 108–09]. The rhetoric of development, possibilities and impossibilities of cyberfeminism open up spaces for accommodating technology [6, pp. 616].

12.2 Communication and Consciousness

The complex digital environment accepts interconnectedness between communication and consciousness. This untethers conventional forms of subjectivity. Due to interwoven links and mediatization changes in intersubjective relations between culture and society is quite common to posthumanist condition [13, p. 244]. "Digital communities index the internal, the external, and the luminal structuring of social worlds in dialogue with realities as defined by media" (p. 245). Virtuality, therefore, has the potential to generate subjectivities in post-digital social life. Sadie Plant's exploration of flexibility and mobility of nomadic subject of the Web explains new virtual communities capable of freedom [15, p. 435]. Nouraie-Simone believes that the absence of physical body in electronic space offers liberating effect on repressed social identities. The allure of disembodiment in cyberspace suggests subversion to gender oppression [4, pp. 111–12]. Communications in cyberspace recraft bodies. The technological tools enforce new relations for women worldwide. The social relations of science and technology reconstruct postmodern collective and personal self. The permeable boundary of body and cyborg permits disassembling and reassembling of self. All heterogeneity disappears in disassembly, reassembly and exchange [9, pp. 102–03]. Cyberfeminists believe that women should use cybertechnologies to create space for empowering themselves. Nancy Paterson observes:

> Cyberfeminism as a philosophy has the potential to create a poetic, passionate, political identity and unity without relying on the logic and language of exclusion or appropriation. It offers a route for reconstructing feminist politics through theory and practise with a focus on the implications of new technology rather than on factors which are divisive. [7, p. 617]

As more women participate in power fields through communication in cyberspace, it assures egalitarian world. Many movements on the social media have voiced gender inequalities and violence. HeforShe, a solidarity campaign initiated by UN, encourages men and boys to be agents of change (act against gender inequalities) for engendering equality. This campaign (2014) was launched on social media. #YesAllWome (2014), #WhyIStayed, #AskHerMore, Bell Bajao, #METOO and Equal Pay Today are several campaigns that transcend borders and reach feminists worldwide. Social media sites have generated awareness about gender issues through dedicated pages. The social media feminism is also termed as the fourth wave of feminism. Though this wave is in its nascent stage, the effectiveness and expanse of cyber activism fulfils the tenets of global social movement. Twitter has also been the engine for the growth of campaigns for gender equality and women empowerment.

12.2.1 Women, Patriarchy and Digital Space

'The body implies mortality, vulnerability, agency: the skin and the flesh expose us to gaze of others . . . or the site where 'doing' and 'being done to' become equivocal', observes Judith Butler [3, pp. 21]. The emergence of the Arab Women's Solidarity Association United (AWSA United) in cyberspace in 1999 enriched activism for women's rights in the Arab world. This kind of pluralistic, transnational and woman's advocacy fosters collective identity and strengthens solidarity. The endeavour of Arab women in diaspora connected six continents to express their dissent against socio-cultural apathy as well as fetters, oppressive patriarchy and repressive government policies. Just as collectivization of identity in physical space is significant for activism, similar cultural characteristics connect the victims of marginalization to construct identities in cyberspace. The proliferation of inexpensive and extensive communication among transnational participants in online environments creates non-governmental intervention in political structure. There is no denying the fact that online environments can fragment identities. Cyber-snooping and data-mining threaten identities but internet provides safety from direct violence and greater geographical freedom. Cybercampaigns, 2011 Arab Spring, 2009 Iranian revolution and hashtag movements demonstrate that cyberspace is a contested space for gender power relations. Cyberfeminism allows postmodern fragmented subjectivities, minorities and women a safe space. It aims to overcome power differences that exclude women from masculine digital space [17, pp. 81–83]. Cyberfeminism provides space to women for redefining gender roles. This alternative discursive space also advances activism and empowers women in the public sphere. Identity exploration takes place through carnivalesque forms of protest in a competitive media environment. Claimed spaces facilitate identity exploration and expression of freedom [2, pp. 244–245]. Dale Spender encourages women to appropriate computer networking [17, pp. 192]. She believes that women must embrace the potentially liberating space for egalitarian principles [pp. 230].

12.2.2 Disembodied Space and Identity of Women

The internet enables women to negotiate identity in disembodied space which is liberatory. Disembodiment, absence of the body online or concealment of identity, which grants greater freedom to women in cyberspace is net utopianism that carries material inequalities. Non-virtual realities of gender, class, race and other cultural variables impact technological experiences. "Virtual interactions are inevitably shaped by, and grounded in, the social, bodily and cultural experience of users" [12, p. 200]. Interaction on internet disrupts ordinary ways of shaping selves and identities because body is represented through words, images, codes and symbols. Women can transgress gender roles, invent selves and create new knowledge. Web pages offer opportunities to women for knowledge creation that might be political aspect of the body/gender. In the recent times internet has become a medium for affirming norms of femininity, individualism and consumerism. As internet generates awareness and agency in relation to illness also, it is "a site where women with breast cancer not only receive information about illness but also compose and circulate their own stories of breast cancer" [14, p. 33]. Women with breast cancer, on the contrary, face a number of challenges in controlling their own definitions and framing their own experiences. The high tech masculine definitions, multiple forms of medical surveillance and social messages in relation to breast cancer detection place burden on women managing the disease. Media and medical industry have framed breast cancer in ways that pressurize women to look normal, erase signs of illness and re-beautify themselves. "Look Good, Feel Better" (LFGB) campaign, co-sponsored by the American Cancer Society and the cosmetic industry, aims to teach women recovering from chemotherapy ways to camouflage signs of illness (pp. 37–38).

12.3 Activism in Cyberspace

"Cyberfeminisms have set out to challenge the male centred culture of internet and to imprint their own models of open and accessible computer-mediated communication onto the new technologies" [11, p. 36]. Information technologies dedicated to progressive social ends create avenues for contemporary feminist interventions. Feminist listserv emails in the 1990s facilitated women discussions and created an online women community for activism. Over the years potential of cyberspace to create egalitarian community has been largely undermined by the patriarchal values and virtual world informed by cultural contexts. As a backlash to Susan G. Komen's decision to remove monetary support of Planned Parenthood in 2014, feminist activist voices generated and consolidated online spurred a change [18, pp. 22–24]. Despite the expansion and appeal of e-movements, cyberactivism for women empowerment confronts several challenges. These movements have been the prerogative of educated women of upper class or professionals. High female illiteracy and unfamiliarity with foreign languages (and English) restrict women's access to web. Limited connectivity of internet is another challenge to cyberactivism of women. Many a times direct or indirect denial of access to certain websites also hinders the e-movements for emancipation of women. Undue support of *fatwa* or rule that forbids women from using internet unless accompanied by a knowledgeable *mahrem* (male guardian) silences the marginalized women. Domination of politicized perspectives and biased media can severely fragment identities [17, pp. 85–86]. The egalitarian achievements in the last few decades do not ascertain equal privacy benefits to women in the realm of cyberspace. Women do not enjoy the same level and types of privacy in cyberspace as men do. The reason for this disparity ranges from complex gendered social norms to perceptions of women as inferiors, ancillaries, soft targets and more accountable for their private conduct. Notably, the norms in real world and cyberworld are the same. In such conditions women-centric perspectives on privacy in cyberspace become vital for engendering equality. Cyberspace is metaphorical for human community as it binds together by networks of relationships, kinship, social and professional. Therefore, meaningful forms of privacy and dignity in cyberspace must be provided to women [1, pp. 1177–1179]. Privacy of women in cyberspace is in peril as many instances of sexual harassment, prying, eavesdropping and emotional injury mark the cyberworld. Women operating in cyberspace are accessible with their names, gender and personal traits. Though a few protection options are available in the cyberspace and women do engage in selective concealment, they remain vulnerable to data collectors and vulgar distractions (pp. 1185–1186). The live broadcast of double mastectomy of Patti on October 20, 1999 in Pennsylvania signals the end of medical privacy. Patti's decision to Webcast Mastectomy also implies end of shame and secrecy regarding breast cancer. Another middle-aged woman, Elizabeth Begat Sean, gave birth on the internet/network to educate others. These instances are a blow to privacy (pp. 1188–1189). Hence, the effective use of internet to increase women's empowerment may be overshadowed by commercialization and consumerism [12, p. 200].

12.4 Violence Against Women in Cyberspace

Individuals forge new relationships, construct multiple cyberidentities and transgress the boundaries of real or physical world. The anonymity and forays into new selfhood also allow perpetrators of violence online spaces to harass women. The sophisticated technologies that aid morphing and construction of fake images propelled cycles of victimization. There is an urgent need to frame policies and laws that protect women from cyberviolence. Policies to address online violence with adequate provisions to check infringement of women's privacy must be the priority of states/nations. Anita Gurumurthy and Niveditha Menon, in the article 'Violence against Women via Cyberspace', comment:

> The slippages between the private and the public that have come to fundamentally reconfigure the spatialities of social transactions and communications characterizing contemporary life dislodge the basic conceptions of feminist thought around the public and private. (10, p. 20)

12.5 Conclusion

Digital dangers arise because the private communication on the internet takes place on platforms that are public in nature. The absence of global framework for resolution of cybercrime requires attention of policymakers. Policies to promote appropriate technologies that create secure and empowering spaces become imperative. Furthermore, the corporatized governance regimes in cyberspace define the rules of networking space. In order to seek optimal solution to challenges faced by cyberfeminism, other feminists like Judy Wajcman and Montserrat Boix propose to reinforce social mechanisms necessary for inclusion of women in the use of technology. Wajcman turns to technofeminism. Technofeminism situates cyberfeminism in social reality that impedes women's access to new technologies. Angustias Bertomeu and Montserrat Boix advance from cyberfeminism to social version of cyberfeminism. In the era of emerging institutional order that is scaffolded by ICTs, the network society creates new exclusions. Feminist analytical frameworks and engagement with policies will greatly reshape dominant taxonomies and embedded experiences of the globalized information society to empower women in cyberspace.

References

1. Allen, A. L. (2000, May). Gender and privacy in cyberspace. *Stanford Law Review, 52*(5), 1175–1200. https://www.jstor.org/stable/1229512
2. Asenbaum, H. (2020). Making a difference: Toward a feminist democratic theory in the digital age. *Politics and Gender, 16*, 230–257.
3. Butler, J. (2004). *Undoing gender*. Routledge.
4. Daniels, J. (2009, Spring–Summer). Rethinking cyberfeminism(s): Race, gender, and embodiment. *Women's Studies Quarterly, 37*(1/2 Technologies), 101–124. https://www.jstor.org/stable/27655141
5. Everett, A. (2004, Autumn). On cyberfeminism and cyberwomanism: High-tech mediations of feminism's discontent. *Signs, 30*(1), 1278–1286. https://www.jstor.org/stable/10.1086/422235.
6. Gajjala, R. (1999). 'Third World' perspectives on cyberfeminism. *Development in Practice, 9*(5), 616–619. https://www.jstor.org/stable/23317590
7. Gajjala, R. (2003). South Asian digital diasporas and cyberfeminist webs: Negotiating globalization, nation, gender, and information technology design. *Contemporary South Asia, 12*(1), 41–56.
8. Gill, R. (2005, March). Review: *Technofeminism. Science as Culture, 14*(1), 97–101.
9. Haraway, D. (1985). A manifesto for cyborgs: Science, technology, and socialist feminism in the 1980s. *Socialist Review, 80*, 65–108.

10. Gurumurthy, A., & Menon, N. (2009). Violence against women via cyberspace. *Economic & Political Weekly, 44*(40), 19–21.
11. Luckman, S. (1999). (En)gendering the digital body: Feminism and the internet. *Hecate, 25*(2), 36–47.
12. Madge, C., & O'Connor, H. (2006). Parenting gone wired: Empowerment of new mothers on the internet. *Social and Cultural Geography, 7*(2), 199–220.
13. Pahwa, S., & Lewis, W. W. (2019, November 7). Reterritorializing digital performance from south to north. *International Journal of Performance Arts and Digital Media, 15*(3), 243–248.
14. Pitts, V. (2004). Illness and internet empowerment: Writing and reading breast cancer in cyberspace. *Health: An Interdisciplinary Journal for the Social Study of Health, Illness and Medicine, 8*(1), 33–59.
15. Puente, S. N. (2008). From cyberfeminism to technofeminism: From an essentialist perspective to social cyberfeminism in certain feminist practices in Spain. *Women's Studies International Forum, 31*, 434–440.
16. Spender, D. (1995). *Nattering on the net: Women, power, and cyberspace*. University of Toronto Press.
17. Stephan, R. (2013, Winter). Creating solidarity in cyberspace: The case of Arab Women's Solidarity Association United. *Journal of Middle East Women's Studies, 9*(1), 81–109. https://www.jstor.org/stable/10.2979/jmiddeastwomstud.9.1.81?seq=1
18. Talay, S. (2013). Feminism and social media: The dilemma of pro-ana websites. *Proceedings of GREAT Day, 1*(4), 21–35.

Chapter 13
Enhancing Digital Leadership Direction: Insight into Empowering Gender Violence Prevention

Miftachul Huda, Syamsul Arifin, Abdul Halim Ali, Abu Zarrin Selamat, Mohd Hairy Ibrahim, Azmil Hashim, Nor Kalsum Mohd Isa, and Zaizul Ab Rahman

13.1 Introduction

Amidst the advancement of technology with its features across the human life, the tendency and the way of human life goes into the digital scenario. Such development is now entering to all cross human life including the organisation [1]. It is clearly alerted that the wide range of strategic attempts to build the digital leadership skills is considered to give a direction on organising the transformation from the physical to virtual basis [1, 2]. The ultimate aims of such development are strategically configured with arranging the interaction and interrelation within the teamwork. The further enhancement on ensuring the digital-utilised practice requires a well-balanced scenario with competitive and cooperative empowerment to support the organisation direction [3]. The ultimate point of driving the digital shift in organising the commitment in the organisation context refers to enhance both efficiency and effectivity to sustain the operation within the guide way of planning strategy. On this view, the enlargement on ensuring the process to run well amidst the systemic design should be taken into consideration through elaborating

M. Huda (✉) · A. Z. Selamat · M. H. Ibrahim · A. Hashim · N. K. Mohd Isa
Faculty of Human Sciences, Universiti Pendidikan Sultan Idris,
Tanjung Malim, Perak, Malaysia
e-mail: miftachul@fsk.upsi.edu.my

S. Arifin
Universitas Islam Negeri Mataram, Mataram, Indonesia

A. H. Ali
Faculty of Language and Communication, Universiti Pendidikan Sultan Idris,
Tanjung Malim, Perak, Malaysia

Z. Ab Rahman
Universiti Kebangsaan Malaysia, Bangi, Selangor, Malaysia

D. Mishra et al. (eds.), *Communication Technology and Gender Violence*, Signals and Communication Technology, https://doi.org/10.1007/978-3-031-45237-6_13

the digital leadership enhancement. With this regard, the strategic arrangement is necessary to drive the process enhancing the leadership direction amidst the digital age.

In line with having the commitment to continue leadership arrangement within the track path, it is necessary to highlight the strategic direction on how to achieve amidst the digital age. It indicated that the necessary enhancement to comprehend the shifting paradigm from physical-traditional basis to virtual online scenario played a role in driving the administration process. The existing arrangement on building the way to run process as the actualisation of engaging the leadership should be integrated with technological planning [4]. On this view, the particular circumstance in conditioning the proper manner fitted to the landscape in responding to the gender violence for instance requires a continued development of responsible awareness on ensuring the entire process from the beginning to the end. At this point, the achievement on driving the leadership amidst the digital age is obviously being the new trends to advance the skills and abilities mainly in restoring the decision-making process amidst the gender violence potentials [5]. As a result, the awareness on looking at the target as the main goal in directing the administration process is required to enhance the responsibility management as a responsive determination to the violence of gender potentials.

In particular, the attempts on actualising the benchmarking arrangement through the critical identification on digital leadership style are required for further study elaboration. There were a number of studies on advancing the digital leadership from the perspectives of approaching system and strategy and also transformation process amidst the five digital revolution industry [6]. Among such varied scholarly works, there has been a lack of academic attention mainly on addressing the real advancement on looking into detail about the digital transformation on directing the leadership enhancement in the attempts to combat the potentials of gender violence. As such, this paper aims to have a critical look at the way on how to drive and achieve the leadership strategy and approach amidst the digital information transformation as an effort to tackle the gender violence prevention arrangement scenario [7, 8]. The further elaboration was made on examining the strategic arrangement on professional skills in underlying the way of digital skills to support the facility on leadership practice.

13.2 Literature Review

13.2.1 Digital Leadership and Responsibility in Organisation Context

The essence of achieving the strategic use amongst the company in running their digital assets for the business purpose could begin with addressing the phase of individual and levels. The strategic approach and skills on how to drive the digital arrangement amidst its process for administration purpose especially are required to organise the circumstances to utilise the achievement of the business goals [9]. Moreover, it is necessary to point out disseminating the requirement procedure as the responsibility to maintain the organisation through the technological advancement reflected into the strategic skills. In particular, the features of digital literacy in order to have a critical look at the key feature in bringing into the organisational trends should begin with adopting the strategic balance between social orientation and individual context [10, 11]. At this point of view, the delivering process of technology adoption into sustaining the leadership practice might become the opportunities in considering the approach with a proper manner [12]. The proper management in responding to the current trends of leadership should bring along with the digital arrangement skills as the form of responsible awareness in fulfilling the digital online transmission age. With this regard, the strategic effort on advancing the effective instrumental design through elaborating the resources could provide the potentials in resulting the beneficial value to gain amidst the digital circumstance. It indicated that the proper arrangement to organise the leadership practice has to bring both ability and capability on driving process within the clear vision [2, 13].

In addition, the special attention should be paid to the achievement of the goal in order to grasp the essence of what is to be done in accordance with the task and duty. It was pointed out that digital leadership emerges from the digital stewardship agreement, which supports the contextual approach of an organisational phase to be continued with an effective tool [14, 15]. In such a phase of expansion of the digital literacy approach, the leadership style that accompanies the successful implementation should assume sufficient commitment to accept the wide range of benefits from the alignment of digital assets. On this view, the digital approach, which enables the leadership commitment to teamwork, could focus on particular attention on achieving a balance between competition and collaboration [16]. In maintaining this simultaneous approach, the digital-based competitive performance for working in a flexible situation is an extension of this digital leadership to address in detail the way of technological situation that could lead to support the business process. Mutual support through well-prepared norms in the company requires a multidimensional approach to respond appropriately in line with the principle guideline in maintaining the safety concern and away from the harmful impact [13, 17]. In terms of enhancing the strategic line on consolidating the digital leadership, the project arrangement needs to be consolidated in designing the plan in order to achieve the potential of achievement plan.

In further, the strategic way of the digital leadership style in assisting to obtain sufficient comprehension is potentially enhanced to give insight into providing the role in playing the proper arrangement amidst the digital circumstance referring to the basic principle scenario [18, 19]. The significant contribution in driving the pathway of the digital environment needs to appropriately deal with continuing the strategic approach of direction and application in the context of leadership management [20]. With this regard, the significant point of having the proper understanding mainly on driving the digital transformation should be incorporated with arranging the planning strategy. The strategic role in helping to drive in delivering the information sources into the organisation members is required to have the proper manner on developing the design on the digital leadership [21]. Such advancement should come up with building the committed awareness as an essential platform in helping to develop in continuing an effective use of digital environment in the context of organisation process and practices. Moreover, the strategic pathway in facilitating the digital facilities in underlying the leadership skills needs to perform the particular transmission on solving abilities. As an essential platform to enhance the organisation achievement, the proper comprehension to look into detail about the contextual circumstance could help in disseminating the exact point on the intrinsic and extrinsic meaning [22]. It is sure to give a clear picture on responding to the current circumstance that the strategic attempts on consolidating the digital use with a proper manner to contribute into helping the process, practices and procedures in solving the issues amidst the organisation.

13.2.2 Digital Leadership as Organisation Stability for Gender Violence Prevention

The aim of digital leadership in helping to make organisation stable is positively linked into the proper adoption of digital technologies with allowing the transmission process followed with the structure of violence prevention attempts. The strategic value in organising the digital skills with its further development requires a proper arrangement in ensuring its effectivity to contribute the stability on advancing the violence prevention scenario [23, 24]. Both services and products are strategically developed further in line with the wide range of having the ability to transmit the process together with its procedure and practices [25]. It is important to transfer the knowledge inquiry about the proper adoption of firm's digital facilities in helping obtain the business goals as the target orientation. With this regard, the ability in running the digital facilities in providing the organisation stability should come up with organising the transmission process to advance in operating process amidst the teamwork. As a result, the strategic approach in maintaining the digital circumstance needs to consolidate the procedure in considering the process in enabling the application strategy on the technology adoption [26]. On this view, the strategic attempts to properly arrange the development process together with its practical implementation require a clear understanding in comprehending the organisation stability to prevent the gender violence. In the attempts to obtain digital leadership, the continued effort to build the digital environment is required to have a substantive commitment in fulfilling the organisation stability to prevent the potentials of gender violence.

In addition, the number of distinctive features provided by the digital technologies in helping to disseminate the information is consequently giving the significant contributions in achieving the main target with following the current circumstances. With this regard, the potentials of digital leadership in helping to run the administration process until getting the organisation stability would play a significant role in ensuring the effectivity and efficiency to achieve the gender violence prevention [27]. In the attempts to concisely gather the organisation society member, comprehending the context in adopting the digital environment requires a strategic approach in facilitating the business process with its instrumental tool to enhance the work productivity performance with the efficient strategy [28, 29]. Moreover, the consistency in continuing such process through adopting the digital circumstance should be taken into consideration in building the strategic approach combined to the properly instrumental tool to help improve the work achievement. The digital facilities in providing the efficient service in the leadership aspects are being the starting point to continue the strategic accomplishment with having the effective and efficient platform [30]. As a result, the making stable of organisation properly from the up to down basis needs to come up with ensuring the strategic arrangement in managing the digital leadership skills. It means that the way to interpret the extensive point of information might potentially contribute into building the mutual line to assist the entire process of organisation pathway in responding and combatting gender violence.

In further, the particular attention should be paid in governing the process through building the achievement plan directed to contribute into carrying out the goal orientation. In the attempts on ensuring the goal transmission to make balance between organisation stability and gender violence prevention, the strategic accomplishment should start with the prominent essence of enhancing the digital leadership in helping transmit the process and practices [1]. On this view, the strategic attempts to apply for the proper application through disseminating both services and facilities are required to have a concise arrangement to obtain the prominent expansion to give the pathway of digital technologies tools for gender violence prevention scenario in line with the targetted plan arrangement [31, 32]. In facilitating the organisation platform to achieve stability among the individuals within the team, the application strategy of digital technologies to continue in ensuring the consistency to support the entire process of management and administration needs to follow the proper arrangement [33]. In order to make such application acceptable and achievable in line with the goal orientation, attempts on continuing the phase in underlying the practising process amidst the virtual environment are strategically required to organise in building the stability orientation. The attempts on having adopted the virtual orientation as the digital platform are strategically required to follow the cloud-basis in delivering an important instrument to ensure the digital information.

13.2.3 Gender Violence in Digital Organisation Environment Context

The existence of violence has been widely emerged into the number of forms one of which is gender violence. Its appearance is now going into the digital age where the varied forms were also reflecting into such examples. It includes as the example violence, harassment, and discrimination which might happen in both virtual and physical basis [3]. Moreover, such occurrence has been potentially leading to the targeted individuals to result in the barriers on the programs and systems within the organisation for instance. Attempts to ensure the effectivity of digital technologies in helping the number of individuals in both online and offline scenarios need to gather occur online and target based on their gender. Amidst the emerging trends and the rise of digital communication platforms and social media, the existence of Internet has become a space where gender-based violence can occur and perpetuate [4]. The number of digital violence could be viewed into the following. Those are online harassment and cyberbullying in which the women and marginalised genders often experience harassment, threats, and bullying online. In particular, the existence of social networking site (SNS) as the media platforms is also clearly seen into the number of examples, such as email, messaging apps, and online gaming [5]. The potentials of such point might range from derogatory comments which lead to the hate speech, and also stalking which refers to sharing and publishing the individuals' private information without consent.

In addition, the varied forms of gender violence might come from the revenge porn which points out disseminating the non-consensual sharing of explicit images or even online videos usually made by the former partners' acquaintances. As such, all these gender violence refers to the main point of intention to shame together with giving the harmful impact to the represented person [6]. The wide range of consequences includes the severe emotional, psychological, and social aspects imposed to the victims. The other practices leading to the forms involve coercing or blackmailing individuals where such behaviour contributes to a hostile environment and perpetuates gender-based discrimination and inequality [9]. The form of abuse and violence could also be found in the number of following aspects such as deception, harassment, stalking, or even physical violence against their individuals or partners through this arranged scenario orientation [34, 35]. The occurrence of such discrimination in digital spaces is potentially seen into the gender discrimination through the online forums scenario, in which the groups discussion circumstance amidst the social media platforms impacted to both men and women to gather the gender minorities. As such, the point of this motive may face an exclusion, silencing, and also dismissive treatment due to their gender [12]. The over part of gender violence in the digital environment is also seen into the form of cyberstalking which refers to the attempts to stalking behaviour. In particular, the wider extension on adopting the digital facilities with the purpose of handling the gender violence is required to have the sufficient monitoring system together with tracking the victims in online basis.

In line with the number of cyber violence including the extent of persistent message, the false rumours and also the invasive surveillance, the need to have a sufficient commitment in addressing such issues mainly in the online basis should be taken into consideration in handling the gender violence. It is important to note that the digital age requires the playfully key performance in underlying the multifaceted approach including both individuals and communities' basis [36, 37]. With this regard, the essence of providing the featured characteristics of digital technology for instance requires to work together in line with the targeted plan [16]. Both companies and policymakers played a role in ensuring the actual implementation reflected into promoting digital literacy together with its awareness. With having the proper manner in supporting the services, the gender violence in the digital age could be handled by giving the continued support in ensuring the victims to have the secure assurance. In particular, the strategic attempts to foster the circumstance on recovery engagement need to reflect into the direction mainly in providing the digital infrastructure to transmit the process in facilitating the digital online features. It is strategically enhanced to give insight into encouraging the initiative to build the support services to advance the commitment to have the respectful awareness together with digital responsibility [13]. At this point, both accessibility and adaptability are required to work together with collaborating the process of managing the digital violence. The need to work in transmitting the knowledge comprehension into building the actual implementation among the organisation membership should bring along with expanding the digital circumstance [38, 39]. The particular contribution to expand the digital accessibility and adaptability is required to enhance the certain information content within the digital collaboration enhancement. As such, the digital access in contributing to develop the action process to organise the challenges to commit with having the actualisation of solving initiative.

13.3 Methodology

This paper aims to examine the forms of digital facilities in underlying the process of leadership process and practices. The critical review has been conducted from the literature on the topic of 'digital leadership to sustain the organisation arrangement'. Through applying for such literature, the analysis was made through gathering review on the topic of professional soft skills on digital leadership competence. As a result, the extent of determining the sufficient comprehension on expanding the main point should bring along with building the significant aspect in facilitating the leadership quality. It is necessary to maintain the strategic empowerment in enhancing the digital-based skills to sustain the cooperative and competitive quality in managing the organisation partnership. Through enhancing the strength on organising the quality on leadership stability, the essence of looking at the gender violence prevention through employing the proper strategy refers to digital sustainability to achieve the organisation quality. Both approach and method in reviewing the recent literatures should be initiated through enhancing the strategic pattern on monitoring the quality with its stability. Moreover, attempts to gain the continued engagement to achieve the strategic alignment through advancing the skills to develop both competitive and cooperative aspects are being the standing point to build an organisation stability. As such, the strategic performance to handle the prevention of gender violence needs to have a sufficient inquiry process to adopt and adapt digital skills organised with the creative, competitive and cooperative empowerment. Through searching for the literature from peer-reviewed articles of journals, proceedings, chapters and books, the strategic enhancement was employed by the key features of elements related to the field. The ultimate point in organising the digital leadership for organisation stability was approached through the varied initiatives including interpreting, conceptualising, and modelling in order to deal with a critical discourse on achieving the main point to battle the gender violence prevention.

13.4 Analysis and Discussion

13.4.1 Enhancing Digital Organisation Stability for Gender Violence Prevention

Attempts to continue the systematic design to accelerating the digital facilities in enhancing the stability refer to empower the initiatives on expanding the digital organisation within the teamwork partnership. In order to ensure the driving strategy in managing the stability of organisation, the ultimate point aims to help the extent of digital leadership commitment to strategize the direction in driving the specific marketplace level [20]. As a result, the arrangement to manage the driving process in monitoring the organisation to sustain in achieving the stability of preventing gender violence. The initiative of empowering the organisation stability is aimed at helping to achieve successfully in taking an advantage from digital initiative arrangement. Moreover, the strategic approach in building the organisation arrangement through digital facilities needs to gain the essence of strategic cooperation to help enhance the competitive advantage [21]. Attempts to explore the strategic concern in the pathway of digital leadership are required to sustain the proper arrangement on the way of digital online quality on the information transmission. Moreover, the digital organisation features and facilities should do with empowering the technology adoption in contributing to the process and practices. With this regard, the key features on building the valuable insight to sustain the consistency in implementing the organisation partnership should be balanced with adapting and adopting the digital-based leadership [25]. In particular, the strategic practice on monitoring the online-oriented structure refers to give insights into building the committed awareness in enhancing the responsibility and responsivity [40, 41]. These two are being the additional key indicators to help arrange the planning strategy to maintain the strategic process and practices in developing the organisation society. As such, the need to continue in adapting the digital skills should be taken into consideration in helping to empower the strategic alignment for leadership organisation.

In addition, the attempts on building the required practices should come up enhancing the initial knowledge as the systematic process in underlying the extent of changing organisation orientation. In the attempts to build the digital leadership, the clear comprehension needs to enhance in continuing process in supporting the knowledge inquiry process in order to sustain the organisation stability [26]. In particular, the acculturation in designing the systematic coordination in engaging the cooperative empowerment needs to have the sufficient understanding in sustaining the successful achievement of organisation stability. With this regard, the cooperation in addressing the significant aspect in taking responsibility should come up with building the awareness on having the digital responsibility engagement [22, 42]. Attempts to explore the strategic attempts on building the process of adapting and adopting the online sources available online are strategically required to enhance the organisation sustainability. The order to obtain an efficient organisation arrangement should do with enhancing the strategic approach in bringing the continued support of digital information achievability [27]. Through the strategic arrangement in building the active promotion in adopting the continued support, the extensive point of digital information is potentially arranged into sustaining the generation of the company achievement plan might be enhanced across the various ecosystems in the digital partnership [43, 44]. In particular, the continued support of digital environment for the organisations stability requires to promote the value of adopting the digital leadership in placing the essence of creativity, communication and willingness in exploring the strategic approach of addressing business contemplation.

In further, the strategic way of driving the digital information can be used to successfully address the running process of organisation in facing the challenges mainly in the pandemic age [30]. Moreover, the running process of business project arrangement is required to deal with engaging digital facilities advanced through the technical operations in achieving the work performance planning and strategy. At this point, adapting the digital facilities in order to help constructing the strategic planning requires to engage the systematic determination to advance the process and procedure of effective digital leadership [33, 45]. In terms of the initiative to ensure the arrangement in facilitating the digital organisation, attempts to enhance in enabling the organisation should bring along with making the timeline arrangement to organise in running the administration purpose [1]. Moreover, the essence of digital applications to underlie the common use on transmitting both services and products continued to support in expanding the digital accessibility. The continued support in carrying out ensuring the strategic applications combined with technical operations in maintaining the organisation stability amidst the digital environment. With this regard, the need to advance the digital innovation should be made in underlying the process of leadership together with focusing on enhancing the organisation quality [3]. On this view, the procedure in monitoring the strategic arrangement as the driving pathway of taking an advantage aims at giving an insight into managing the procedural operation to achieve the organisation stability as the main strategic initiative empowerment [46, 47]. As such, the valuable insights in advancing the company's digital assets should come up with enhancing the individual phase where the sufficient technical operations need to gather in empowering the digital partnership in underlying the leadership process. In the attempts to continue working with ensuring the responsible commitment, the proper execution in building the digital leadership is being significantly as the way to contribute into optimising the value feedback through knowledge comprehension and also digital information.

13.4.2 Accessing Digital Information-Based Organisation Stability for Gender Violence Prevention

The need to enhance the digital leadership in responding to the gender violence requires the strategic attempts in empowering the key point of taking advantage in organisation context. Moreover, the commitment to take the advantage given through the digital information refers to enhance the sustainable development initiative in order to achieve the positive value for organisation stability [4]. Through enhancing the digital information, the empowerment of managing the organisation structure could be managed through having the instructional design sufficiently in line with advancing the organisation stability [41, 48]. The essence of accessing the digital information basis aims to give an insightful value in enabling to monitor the entire process and procedure. With this regard, the strategic enhancement to provide what to do properly amidst the organisation arrangement could facilitate the sufficient instrumental tool in order to continue both products and services [5]. On this view, the supply process organised through managing the proper arrangement to achieve the organisation stability should bring along with adjusting the allocation of timely basis with an effective transmission of information delivery process. The proper requirement in bringing the condition is required to drive into the digital application amidst the teamwork commitment in leading to support the wellness as the fundamental element to drive a key indicator for digital society partnership scenario [34, 49]. In the attempts to enhance the running process for the organisation arrangement purpose, the strategic attempts to maintain the proper arrangement should be taken into consideration in performing the appropriate coordination [6]. As such, the active acceleration in determining the various approaches in implementing the digital leadership could be the standing point on helping to monitoring the process in serving as underlying the achievement of organisation stability, in that the purpose aims at enhancing the target to manage the gender violence anticipation and prevention scenario as the main contribution.

In addition, the proper arrangement as the way to monitor the digital leadership needs to ensure the dissemination process of information sustained into building the aspects of what can help in increasing the advantage and profit for the organisation stability should come up with enhancing the maintenance phase [9]. One of the ultimate points of this phase refers to give a better service into the customers through the digital instrumental tools, in enabling to achieve an increasingly satisfied quality. Moreover, the wider approach of digital information accessibility turned into maintaining the loyal customers for the organisation stability would increase the visibility of quality amongst the customers receiving the service [12]. As such, the strategic digital tools could be developed further in serving the proper benefit and giving the facilities in supporting the entire process of transaction, social interaction and also education transmission procedure scenario [50, 51]. Moreover, the operational rate in customising the cost with the lower basis mainly in the economic sector refers to organise need to expand the strategic approach of digital leadership in sustaining the administration process. Comprehending the extent of what to do properly should be clearly balanced to carry out using the digital facilities requires to have achieve its accessibility and adaptability [52]. At this point of view, the continued capacity in line with helping to achieve the organisation running through obtaining the administration is significantly endowed with building the initiative of enhancing the digital access oriented transmission process and strategic effort empowerment [53, 54]. In particular, the entire process in pointing out the necessary act of leadership expansion would enable in contributing to advance the strategic arrangement in looking into detail on driving the running process itself [16]. Thus, the leadership style in going further to achieve the vision as strategized in the planning system requires to commit with having the digital skills organised into expanding the overall work achievement plan.

In further, the essence of unique set of having the skills in running the process to driving the outcome is required to provide the attempts on what to do in line with building the organisation stability. Moreover, the strategic expansion on assisting the work achievement mainly in supporting the digital leadership enhancement needs to have a solid consolidation to engage with developing the sufficiently sustainable quality [13]. In the attempts to empower the extensive point of knowing what to do in line with enhancing the leadership development, the important aspects to build the access of digital skill should be accommodated in promoting the gender violence prevention. With this regard, the strategic appointment needs to have a sufficient detail on looking into the capacity and capability in ensuring the extent of an increasingly important pathway of achievement. The point is that enhancing the sustainable information through promoting the operation should do with achieving the particular attention in developing the strategic initiative of digital information for gender violence purpose [20]. As such, the point of having the stability with empowering the mutual line of adopting the technology enhancement should bring along with the significant enhancement to sustain the linked network for communication pathway. Attempts to empower the accessibility of digital information are strategically advanced in building the required practices on advancing the strategic phase of leadership skills with digital adaptation and adoption in supporting the process and procedure of driving the strategic attempts in line with the objective of the agenda [55, 56]. The main point refers to give insights into monitoring the process and practices of digital leadership in supporting the continued arrangement through expanding the communication strategy. On this view, the strategic point in approaching way of building the digital communication pathway needs to gather the wide range of the digital leadership through accelerating the comprehensive effort on understanding the alignment from the top to the bottom orientation.

13.4.3 Adapting Sufficient Digital Sustainability for Organisation Stability with Gender Violence Prevention

The continued development on having the sufficient adaption is aimed at helping to strategize the skills of problem solving. This is important because it might give insight into comprehending the circumstance and situation to face the challenges of both industrial and social. On the companies and industries, they are required to provide the sustainability of accessing the technology and digital facilities in order to assist proceeding the generating process in achieving the better service [21]. With this regard, the arrangement on adapting the digital technologies environment needs to stabilise both capability and capacity to access an important pathway to obtain the better service through engaging the digital skills management in supporting the process on monitoring the preventive action towards the unforseen circumstance [57, 58]. On this view, the particular result is that the strategic point in ensuring the process of digital leadership enhancement in sustaining the organisation stability is required to grip with the essence of technology tools [25]. The point is that attempts to have such proper strategy refer to give insight into bringing the overall work performance [59, 60]. As a result, the essence of teamwork partnership to facilitate point on having the proper guideline should do with managing the strategically proper manner in driving the digital environment. Moreover, the wider context of achieving the balance between strategy and application is required to come up with advancing the workforce initiative to build the organisation teamwork [26]. The particular attention should be paid in enabling the individual and social membership in order to adopting and adapting the digital facilities in supporting the leadership pathway. The exact point on how to operate the organisation process to manage handling the gender violence is starting the proper strategy in comprehending the successful achievement through obtaining both goal and vision. As such, the achievement process and procedure should bring along with building the commitment in continuing the productivity to achieve the future pathway of digital leadership [27]. The strategic goal as noted in the vision needs to clearly enhance sustaining the strategic expansion of having a very clear vision for the future leadership pathway.

In addition, particular attention needs to be given into paying the arrangement of digital leadership enhancement to continue empowering the competency in determining the strategy on assisting the process and practices. In order to achieve this plan, obtaining the strategic practice should begin with enhancing the preparation with having a sufficient adaptation of digital space in underlying the organisation environment space [30]. Attempts to expand the proper manner reflected into the way to transmit the information are supposed to sustaining the planned arrangement to further implement in underlying the order of possible changes in organisation context. With this regard, the strategic plan as arranged in the standard needs to have the actual point of comprehending the clear vision together with coherently applicable strategy [61, 62]. Moreover, the crucial pathway to enhance the initiative of expanding the digital space is needed to have the clear vision with building the agenda in initiating the process of work performance and innovations [1]. Through underlying the innovation procedure, the strategic appointment in contributing to build the digital leadership is required to continue the sustainable development agenda for instance as this might be transmitted into the mutual collaboration between accessibility of technology and stability of organisation [63]. At this point of view, the digital transmission process in advancing the initiative to handle in managing the gender violence should bring along with initiating the leadership pathway.

In further, the strategy implementation with an enthusiastic point could be carried out in advancing the digital technologies in order to enable the process of flexibility in the management. In order to maintain the arrangement with adaptability and stability, considering the strategic approach in building the digital partnership should be clearly and carefully calculated in giving the proper manner in solving the potential risks [3]. With this regard, the proper arrangement in building the mutual line in facilitating the process of improvement needs to adopt both procedure and gateway mainly in the extent of measuring the sufficient digital adaptation amidst the normal platform [64, 65]. In the attempts to enhance the potential skills on enlarging the organisation style, the need to have the strategic transmission on building the detailed plan to achieve the targeted goal should be taken into consideration on assisting the business achievement [4]. It means that the proper understanding on adapting and adopting the digital transformation could be employed with the strategic arrangement on building the platform to ensure the instrumental tool to provide an appropriate facility as the supporting arrangement [66]. Through expanding the complete arrangement in facilitating the organisation procedure, the strategic attempts in providing the digital transformation in resulting the important part in order to supplement the circumstance of organisation pathway would become a tremendous significance in contributing to assist in enhancing the digital leadership practices.

In line with the frequent transmission of digital adaptation and adoption to ensure the prompt solution in responding to the digital challenges, both potentials and risks are the two lines on attracting the application strategy properly amidst the certain condition [35]. The balance between digital ethics and professional skills as an underlying paradigm could give insight into building the reality on driving the practices amidst the virtual context [67, 68]. In order to ensure the commitment to continue the existing leadership applications, the strategic approach to maintain the proper manner into the further achievement on the digital skills is concisely engaged with underlying the leadership process and practices. As such, it is required to pay attention on taking the proportional size in considering the beneficial value in achieving the digital leadership skills [5]. On this view, the transmission process in organising the direction pathway to work in the team with the same time in the attempts to gender violence prevention should gain the performance procedure to achieve the plan as in line with the goal achievement as the key element to drive the strategic direction to involve ethical and professional balance [69, 70]. As such, benchmarking offers a proper arrangement, implementation and assessment towards the attempts in running the digital leadership skills and practices. With this regard, the playful encouragement in managing the attempts in consolidating the digital infrastructure should do with building the particular point in organising the process followed by the practices in order to enhance the delivering systems of strategic use of virtual basis [71]. In order to bring the digital access and facilities, the information delivery process should be taken into consideration in developing the teamwork members to result in having the specific access of the digital environment and infrastructure basis.

13.5 Conclusion

This paper did examine the digital leadership enhancement to sustain the organisation pathway as an effort to the gender violence prevention. As one of the significant elements of driving factors in continuing the organisation stability for gender violence prevention, the digital leadership style arrangement could be delivered in building the strategic coordination through online competence skills. Through driving the digital shift paradigm towards the organisation, operating the process in order for being more responsive and flexible would lead to have the digital professional skills in setting the good process with the expert achievability. This chapter aims to examine the professional soft skills with digital competence in expanding an important element to facilitate gender violence prevention. The finding is focusing on how the driving process could go through an innovative spirit with being more openness and responsiveness towards the operation procedure, mainly on managing the customer relationship. The value of this chapter is arranged to constantly expand the digital competencies transmitted into the employees with the concern of trust and appreciation in such a process. Those are enhancing digital information sustainability and accessibility for organisation stability, advancing sufficient adaptation of digital environment space for organisation stability, continuing digital skills-based goal achievement strategy for organisation stability, strategizing digital technical experts of security for organisation stability, empowering continued support on digital environment with organisation stability for gender violence prevention.

References

1. Miller, K. C. (2021). Hostility toward the press: A synthesis of terms, research, and future directions in examining harassment of journalists. *Digital Journalism*, 1–20.
2. Anshari, M., Almunawar, M. N., Shahrill, M., Wicaksono, D. K., & Huda, M. (2017). Smartphones usage in the classrooms: Learning aid or interference? *Education and Information Technologies, 22*, 3063–3079.
3. Emezue, C. (2020). Digital or digitally delivered responses to domestic and intimate partner violence during COVID-19. *JMIR Public Health and Surveillance, 6*(3), e19831.

4. Laufer, M., Leiser, A., Deacon, B., Perrin de Brichambaut, P., Fecher, B., Kobsda, C., & Hesse, F. (2021). Digital higher education: A divider or bridge builder? Leadership perspectives on edtech in a COVID-19 reality. *International Journal of Educational Technology in Higher Education, 18*, 1–17.
5. Sambasivan, N., Batool, A., Ahmed, N., Matthews, T., Thomas, K., Gaytán-Lugo, L. S., ... & Consolvo, S. (2019). "They don't leave us alone anywhere we go" gender and digital abuse in South Asia. In *Proceedings of the 2019 CHI conference on human factors in computing systems* (pp. 1–14).
6. Storer, H. L., & Rodriguez, M. (2020). #Mapping a movement: Social media, feminist hashtags, and movement building in the digital age. *Journal of Community Practice, 28*(2), 160–176.
7. Huda, M., Ali, A.H., Za, Tabrahni., Roslee, A., Ibrahim, M.H., Anshari, M., Glorino, M. (2023). Understanding of digital ethics for information trust: A critical insight into gender violence anticipation. In Deepanjali Mishra et al. (Eds), Communication technology and gender violence. Springer.
8. Zainuri, A., Sukarno, S., & Huda, M. (2022). Understanding scientific literacy and pedagogy competence: A critical insight into religious integration thinking skills. *Journal of Educational and Social Research, 12*(1), 274–281.
9. Dlamini, N. J. (2021). Gender-based violence, twin pandemic to COVID-19. *Critical Sociology, 47*(4–5), 583–590.
10. Mulyadi, D., Huda, M., & Gusmian, I. (2022). Smart Learning Environment (SLE) in the Fourth Industrial Revolution (IR 4.0). *International Journal of Asian Business and Information Management, 13*(2), 1–23. https://doi.org/10.4018/IJABIM.20220801. https://doi.org/10.4018/IJABIM.287589
11. Huda, M., Sutopo, L., Liberty, F., & Mustafa, M. C. (2022, March). Digital information transparency for cyber security: Critical points in social media trends. In *Advances in Information and Communication: Proceedings of the 2022 Future of Information and Communication Conference (FICC), Volume 2* (pp. 814–831). Springer.
12. Green, L. (2020). Confident, capable and world changing: Teenagers and digital citizenship. *Communication Research and Practice, 6*(1), 6–19.4.
13. Bhatia, A., Fabbri, C., Cerna-Turoff, I., Turner, E., Lokot, M., Warria, A., et al. (2021). Violence against children during the COVID-19 pandemic. *Bulletin of the World Health Organization, 99*(10), 730.
14. Huda, M., Borham, A. H., Almunawar, M. N., Anshari, M., & Ahmad, R. (2023). Strategic role of trust in digital communication: Critical insights into building organizational sustainability. In K. Arai (Ed.), *Proceedings of the Future Technologies Conference (FTC) LNNS 815*, pp. 1–16. Springer.
15. Abdul Aziz, N. A., Mohd Razali, F., & Saari, C. Z. (2022). Penggunaaan Media Sosial dari Perspektif Psiko Spiritual Islam. *Firdaus Journal, 2*(1), 65–75. https://doi.org/10.37134/firdaus.vol2.1.6.2022
16. Peng, J., Nie, Q., & Chen, X. (2023). Managing hospitality employee cyberloafing: The role of empowering leadership. *International Journal of Hospitality Management, 108*, 103349.
17. Huda, M. (2022). Empowering professional and ethical balance in digital record management. *Organizational Cybersecurity Journal: Practice Process and People, 2*(1), 60–73. https://doi.org/10.1108/OCJ-06-2021-0016
18. Jailani, M., Hafidh, N., & Huda, M. (2023). The Influence of Upin And Ipin Animation on Children's Religious Life. *Penamas, 36*(1), 74–90.
19. Ab Rahim, N. M. Z., Saari, Z., Mohamad, A. M., Rashid, M. H., & Mohamad Norzilan, N. I. (2022). Konsep Ulul Albab dalam Al-Quran dan hubungannya dengan pembelajaran kursus Sains, Teknologi dan Manusia di UTM Kuala Lumpur. *Firdaus Journal, 2*(2), 72–78. https://doi.org/10.37134/firdaus.vol2.2.7.2022
20. Carrington, K., Morley, C., Warren, S., Ryan, V., Ball, M., Clarke, J., & Vitis, L. (2021). The impact of COVID-19 pandemic on Australian domestic and family violence services and their clients. *Australian Journal of Social Issues, 56*(4), 539–558.

21. Rauhaus, B. M., Sibila, D., & Johnson, A. F. (2020). Addressing the increase of domestic violence and abuse during the COVID-19 pandemic: A need for empathy, care, and social equity in collaborative planning and responses. *The American Review of Public Administration, 50*(6–7), 668–674.
22. Ramle, M. R., & Huda, M. (2022). Between text and context: Understanding Ḥadīth through Asbab al Wurud. *Religions, 13*(2), 92.
23. Huda, M., Ramli, A. F., Ritonga, M., Rahim, M. M. A., Borham, A. H., Hashim, A., Husain, H., & Isa, N. K. M. (2023a). From digital ethics to digital partnership skills: Driving a safety strategy to expand the digital community? In *Digital transformation for business and society: Contemporary issues and applications in Asia*. Taylor & Francis.
24. Arifin, S., Huda, M., & Mufida, N. H. (2023). Akhlak Karimah Value-Based Integrative Learning Model in Madrasah. *Jurnal Pendidikan Islam, 9*(1), 41–54.
25. Silva, A. F. D., Estrela, F. M., Soares, C. F. S., Magalhães, J. R. F. D., Lima, N. S., Morais, A. C., … & Lima, V. L. D. A. (2020). Marital violence precipitating/intensifying elements during the Covid-19 pandemic. Ciencia & Saude Coletiva, 25, 3475–3480.
26. Reed, L. A., Ward, L. M., Tolman, R. M., Lippman, J. R., & Seabrook, R. C. (2021). The association between stereotypical gender and dating beliefs and digital dating abuse perpetration in adolescent dating relationships. *Journal of Interpersonal Violence, 36*(9–10), NP5561–NP5585.
27. Sharma, A., & Borah, S. B. (2020). Covid-19 and domestic violence: An indirect path to social and economic crisis. *Journal of Family Violence*, 1–7.
28. Maseleno, A., Huda, M., Jasmi, K. A., Basiron, B., Mustari, I., Don, A. G., & bin Ahmad, R. (2019). Hau-Kashyap approach for student's level of expertise. *Egyptian Informatics Journal, 20*(1), 27–32.
29. Rachman, A., Oktoviani, I., & Manurung, P. (2023). Digitization of Zakat and Charity BAZNAS Tangerang City through Crowdfunding Platform tangerangsedekah. id. *Firdaus Journal, 3*(1), 96–106. https://doi.org/10.37134/firdaus.vol3.1.9.2023
30. Fairbairn, J. (2020). Before# MeToo: Violence against women social media work, bystander intervention, and social change. *Societies, 10*(3), 51.
31. Huda, M., & Salem, S. (2022). Maintaining quality family time for children's social intelligence: public educators' beliefs and practices in the pandemic age. In *Children, youth and time* (Vol. 30, pp. 69–90). Emerald Publishing Limited.
32. Qadafi, M., Ulfah, M., Huda, M., & Agustiningsih, N. (2023). Fostering independent learning in early childhood: A case study on montessori pedagogy at PAUD Montessori Futura Indonesia. *Golden Age: Jurnal Ilmiah Tumbuh Kembang Anak Usia Dini, 8*(3), 109–120.
33. Huda, M. (2022). Towards an adaptive ethics on social networking sites (SNS): A critical reflection. *Journal of Information, Communication and Ethics in Society, 20*(2), 273–290.
34. Alwi, A. S. Q., & Ibrahim, R. (2022). Isu terhadap Penggunaan Teknologi Media Digital dalam kalangan guru pelatih jurusan Pendidikan Khas. *Firdaus Journal, 2*(2), 88–93. https://doi.org/10.37134/firdaus.vol2.2.9.2022
35. Huda, M. (2019). Empowering application strategy in the technology adoption: Insights from professional and ethical engagement. *Journal of Science and Technology Policy Management, 10*(1), 172–192. https://doi.org/10.1108/JSTPM-09-2017-0044
36. Khalili, H. M, Rosman, A. S., Mohamed, A. K., & Marni, N. (2021). Digital learning enhancement through social network site (SNS). In *Software Engineering Application in Informatics: Proceedings of 5th Computational Methods in Systems and Software 2021*, vol. 1 (pp. 421–431). Springer.
37. Maimun, M., Huda, M., & Muhammad, M. Strategies for building: A competitive and excellent islamic educational institution. *POTENSIA: Jurnal Kependidikan Islam, 8*(2), 137–153.
38. Huda, M., Ramli, A. F., Ritonga, M., Rahim, M. M. A., Borham, A. H., Hashim, A., Hanafi, H. F., & Isa, N. K. M. (2023b). Towards digital servant leadership for organisational stability: Driving processes in the pandemic age? In *Digital Transformation for Business and Society: Contemporary Issues and Applications in Asia*. Taylor & Francis.

39. Huda, M. (2023). Towards digital access during pandemic age: better learning service or adaptation struggling? *foresight, 25*(1), 82–107. https://doi.org/10.1108/FS-09-2021-0184
40. Zacher, H., & Rudolph, C. W. (2022). Researching employee experiences and behavior in times of crisis: Theoretical and methodological considerations and implications for human resource management. *German Journal of Human Resource Management, 36*(1), 6–31.
41. Sari, D. C., Ali, A. H., Harun, M., Batre, N. M., Hanafi, M. S., Jaludin, Z. Y., et al. (2023). Transformation of artificial intelligence in Islamic Edu with Ulul Albab Value (Global challenge perespective). *Firdaus Journal, 3*(1), 1–9.
42. Huda, M., & Salem, S. (2022). Understanding human behavior development with spirituality: Critical insights into moral flourishing. *Ulumuna, 26*(2), 238–268.
43. Huda, M. (2023). Trust as a key element for quality communication and information management: insights into developing safe cyber-organisational sustainability. *International Journal of Organizational Analysis*. https://doi.org/10.1108/IJOA-12-2022-3532
44. Huda, M. (2023). Digital marketplace for tourism resilience in the pandemic age: voices from budget hotel customers. *International Journal of Organizational Analysis, 31*(1), 149–167. https://doi.org/10.1108/IJOA-10-2021-2987
45. Ropei, A., Huda, M., Alijaya, A., Fadhil, F., & Zulfa, F. (2023). Managing 'Baligh'in four Muslim countries: Egypt, Tunisia, Pakistan, and Indonesia on the minimum age for marriage. *Al-Ahwal: Jurnal Hukum Keluarga Islam, 16*(1), 112–140.
46. Zainuri, A., & Huda, M. (2023). Empowering cooperative teamwork for community service sustainability: Insights from service learning. *Sustainability, 15*(5), 4551.
47. Ramli, A. F., Ramli, N., Huda, M., Asy-Syaimaa'Hussain, L. K., & Ramli, N. I. A. (2022). Critical Investigation on the pandemic from the Islamic perspective. *Afkar: Jurnal Akidah & Pemikiran Islam*, 99-140.
48. Adil, M., & Huda, M. (2023). Understanding responses to worship regulations in the pandemic era: Text data mining analysis in the Indonesian context. *Religions, 14*(4), 549.
49. Yahya, S. F., & Othman, M. A. (2022). Penggunaan video dalam Pengajaran dan Pembelajaran Pendidikan Moral Tingkatan 2. *Firdaus Journal, 2*(2), 94–105. https://doi.org/10.37134/firdaus.vol2.2.10.2022
50. Jamaluddin, N. S. (2023). Applicability of constructivist-oriented teaching in the process of educational transformation in Malaysia. *Firdaus Journal, 3*(1), 51–59. https://doi.org/10.37134/firdaus.vol3.1.6.2023
51. Berlian, Z., & Huda, M. (2022). Reflecting culturally responsive and communicative teaching (CRCT) through partnership commitment. *Education Sciences, 12*(5), 295.
52. Huda, M. (2023). *Between accessibility and adaptability of digital platform: Investigating learners' perspectives on digital learning infrastructure*. Higher Education, Skills and Work-Based Learning.
53. Mohd Nawi, M. Z. (2023). Media variations in education in Malaysia: A 21st century paradigm. *Firdaus Journal, 3*(1), 77–95. https://doi.org/10.37134/firdaus.vol3.1.8.2023
54. Rosa, A. T. R., Pustokhina, I. V., Lydia, E. L., Shankar, K., & Huda, M. (2019). Concept of Electronic Document Management System (EDMS) as an efficient tool for storing document. *Journal of Critical Reviews, 6*(5), 85–90.
55. Kembauw, E., Soekiman, J. F. X. S. E., Lydia, L., Shankar, K., & Huda, M. (2019). Benefits of corporate mentoring for business organization. *Journal of Critical Reviews., 6*(5), 101–106.
56. Gani, A. A., Ibrahim, N., Khaerudin, J., & M., Huda, M., and Maseleno, A. (2019). Exploring multimedia-based active learning pedagogy: An empirical research. *TEST Engineering and Management, 81*, 4311–4321.
57. Muslihudin, M., Ilayaraja, M., Sathesh, K. K., Shankar, K., Jamilah, J., Novitasari, D., Huda, M., Hashim, W., Rudenko, I. V., & Maseleno, A. (2019). Decision support system in Kindergarten selection using TOPSIS method. *International Journal of Recent Technology and Engineering, 8*(1), 3291–3298.
58. Susilowati, T., Manickam, P., Devika, G., Shankar, K., Latifah, L., Muslihudin, M., et al. (2019). Decision support system for determining lecturer scholarships for doctoral study using CBR (Case-based reasoning) method. *International Journal of Recent Technology and Engineering, 8*(1), 3281–3290.

59. Leh, F. C., Anduroh, A., & Huda, M. (2021). Level of knowledge, skills and attitude of trainee teachers on Web 2.0 applications in teaching geography in Malaysia schools. *Heliyon, 7*(12).
60. Huda, M., Gusmian, I., & Mulyo, M. T. (2021). Towards eco-friendly responsibilities: Indonesia field school model cross review. *Journal of Comparative Asian Development (JCAD), 18*(2), 1–12.
61. Fitrian, Y., Huda, M., Muhtar, A., Arifin, A. Y., Musa, N., Teh, M., et al. (2019). Application design for determining suitable cosmetics with the facial skin type using fuzzy logic approach. *Journal of Computational and Theoretical Nanoscience, 16*(5-6), 2153–2158.
62. Abadi, S., Hawi, A., Dacholfany, I., Huda, M., Teh, K. S. M., Walidi, J., et al. (2019). Identification of Sundep, Leafhopper and fungus of paddy by using fuzzy SAW method. *International Journal of Pharmaceutical Research, 11*(1), 695–699.
63. Nurmartiani, E., & Huda, M. (2020). Marketing mix performance and customer relationship in improving trust of indihome customer: A case from West Java Indonesia. *Journal of Critical Reviews, 7*(2), 275–282.
64. Alzaabi, M. A. J., Marni, N., & Huda, M. (2019). Examining Information Accuracy on Social Media: Empirical Evident from United Arab Emirates (UAE). *Journal of Critical Reviews, 536.*
65. Huda, M., Hehsan, A., Basuki, S., Rismayadi, B., Jasmi, K. A., Basiron, B., & Mustari, M. I. (2019). Empowering technology use to promote virtual violence prevention in higher education context. In *Intimacy and developing personal relationships in the virtual world* (pp. 272–291). IGI Global.
66. Hanafi, H. F., Wahab, M. H. A., Selamat, A. Z., Masnan, A. H., & Huda, M. (2021). A systematic review of augmented reality in multimedia learning outcomes in education. In *Intelligent Human Computer Interaction: 12th International Conference, IHCI 2020, Daegu, South Korea, November 24–26, 2020, Proceedings, Part II 12* (pp. 63–72). Springer.
67. Rudolph, C. W., Allan, B., Clark, M., Hertel, G., Hirschi, A., Kunze, F., et al. (2021). Pandemics: Implications for research and practice in industrial and organizational psychology. *Industrial and Organizational Psychology, 14*(1–2), 1–35.
68. Anshari, M., & Almunawar, M. N. (2012). Framework of social customer relationship management in e-health services. *arXiv preprint* arXiv:1207.6179.
69. Huda, M. (2021). Islamic philosophy and ethics of education: Al-Zarnūjī's concept of Ta'ẓīm in his Ta 'līm al-Muta 'allim. *Ulumuna, 25*(2), 399–421.
70. Sukadari, S., & Huda, M. (2021). Culture sustainability through Co-curricular learning program: learning Batik cross review. *Education Sciences, 11*(11), 736.
71. Huda, M., & Teh, K. S. M. (2018). Empowering professional and ethical competence on reflective teaching practice in digital era. In *Mentorship strategies in teacher education* (pp. 136–152). IGI Global.

Chapter 14
Understanding of Digital Ethics for Information Trust: A Critical Insight into Gender Violence Anticipation

Miftachul Huda, Abdul Halim Ali, Tabrani Za, Roslee Ahmad, Abu Zarrin Selamat, Mohd Hairy Ibrahim, Muhammad Anshari, and Moses Glorino

14.1 Introduction

In the last decades, the advancement of technology with its sophisticated features offered the convenient facilities on supporting the human daily needs. The coefficient services through such distinctive features played a significant role in distributing the instruction for the society's demands and aims [8]. Enhancing the initiative to advance the technology facilities has been given in providing the fulfilment initiative to have a potentiality of strategy adoption. One of the examples is through the balance initiative in the attempts to the application strategy through professional and ethical engagement [22]. In terms of the current digital ethics' commitment, the practical phase of focusing on the effort to adopt the technology enhancement should be committed into building the strategic way on driving the online users in a

M. Huda (✉) · A. Z. Selamat · M. H. Ibrahim
Faculty of Human Sciences, Universiti Pendidikan Sultan Idris, Tanjung Malim, Malaysia
e-mail: miftachul@fsk.upsi.edu.my

A. H. Ali
Faculty of Language and Communication, Universiti Pendidikan Sultan Idris, Tanjung Malim, Malaysia

T. Za
Universitas Serambi Mekkah, Kota Banda Aceh, Indonesia

R. Ahmad
Faculty of Management and Leadership, Universiti Sains Islam Malaysia, Nilai, Malaysia

M. Anshari
School of Business, Universiti Brunei Darussalam, Bandar Seri Begawan, Brunei

M. Glorino
Faculty of Humanities, Universitas Airlangga, Surabaya, Indonesia

D. Mishra et al. (eds.), *Communication Technology and Gender Violence*, Signals and Communication Technology, https://doi.org/10.1007/978-3-031-45237-6_14

proper manner arrangement in line with the digital society circumstance and background [49]. As a result, the main concern from such initiative aims at having the assurance to achieve the trust as the information quality.

It is in the current circumstance that the number of violence in the digital age has been gradually emerged with the variety of forms, approaches and motives [9]. The main priority which should be given a portion on obtaining what to do in line with such condition needs to have a careful engagement in responding properly and effectively in line with the digital skills advancement scenario [36]. In order to achieve such point, the appropriate strategy which represented the initiative on combatting such challenges is through taking account into the digital ethics commitment reflected into the practice [12]. Amidst the massive emergence toward the cyber violence, the current approach in responding it with the proper way has to be taken into consideration which could be presented into the committed practice. With this regard, such urgent demand requires to have a commitment in taking account into the digital practice with an ethical engagement arranged to build cyber society conduciveness [23]. As a result, the continued development on ensuring the cyber violence under control should be committed into the practice across the gender consequence. Moreover, the impact would be faced if ignoring such initiative and this requires the real commitment in order to make balance between technological advancement adaptability and ethical and professional applicability [59]. As such, responding to such trends is translated into reconstructing the benefit between the value and the vague as the strategic point to continue the practical stage of having the trust as the key element to sustain organization stability [25].

In addition, the initiative to take a point in considering the positive feedback given by the technological advancement and staying alert to the negative impact possibility requires a clear step accommodated into the human mind and practice. As a result, attempts to actualise the strategic balance of ethical engagement and professional enhancement are required to build the information trust as the quality [70]. With this regard, the process of disseminating the positive feedback in giving insights into the violence of gender throughout such prevention needs to pay attention on reconstructing the application strategy [73]. The number of studies on understanding of digital ethics for information trust with the special focus on developing the initiative for gender violence anticipation has been lack of scholarly attention. As such, the current study aims to focus on developing the digital ethics reflected into the pathway of building the technology adoption in the practical scenario. In particular, the strategic occupation on building the digital ethics-committed practice would give insights into helping to achieve the trust in the information as the quality [74]. The main concern of this study was on reframing the digital ethics' main features in underlying the attempts for information trust in enabling the users to have interaction support amidst the digital circumstance. The current study aims to give the value on building the digital technology enhancement to advance the digital society in driving their purposes, motives, and business orientation with the professional and ethical pathway.

14.2 Literature Review

14.2.1 Defining Digital Ethics for Human Sustainability

The explicit point of digital ethics' core values refers to point out recognising the trust maintenance committed into the online practices [13]. With having a sufficient work employed to result in the core values, the emerging trends of practical stability in recognising the approach of trust compliance in the digital environment both in public and private sphere required a digital ethics' core values consistency in line with the safe cyber orientation [33]. In the way to understand the emerging trends of failing the trust, the potentials are supposed to lead to the human daily life with a critical goal to be pursued [14]. In the attempts to continue a significant transparency in enabling to signify the complex condition, the strategy in fulfilling the market expansion together with the committed digital ethics should do with encompassing the core values on what to do in line with the digital skills integrated in the online practice arrangement [34]. With this regard, the digital ethics actors' capacity amongst regulators, customers, and also the media expect brands are required to have a sufficient openness together with hones commitment in all conditions [15]. It is important to take note in performing the consistent pathway in the capacity of digital users in particular and public society and community in general. As such, the transparency is being an important aspect in delivering the information in order to achieve digital balance and harmony.

In the attempts to comply with the consistency across all aspects of human society, the strategic transparency achievement in underlying the process of promotion, marketing and also products and service distribution would need to have a continued workforce culture and partner relationships within the digital circumstance scenario [26]. The strategic effort to support the human life aspect in communication for instance should begin with consolidating the know-how orientation as the fundamental element in continuing the worldwide dimension to give a positive feedback to the partnership amongst the digital users in using their online social interaction [6]. The disrupting challenges faced by the number of sectors are hardly impacted to enforce the digital information trust amidst the technology expansion. In the attempts to grab the chances in gaining the committed transparency, the continued practice of having digital ethics' core values in mind could become an initial pathway to transmit digital transparency [16]. With this regard, the number of partners such as employees, customers, regulators and investors are to be embedded in organising the the sufficient digital skills to underlie the cyber practice-oriented environment strength [1, 27]. Embedding the organisational value reflected into the ethical principles of digital transparency in online practices is required to apply for demonstrating a commitment to do an appropriate goodness in building the long-term foundation of trust with stakeholders [17]. In this view, the digital ethics' core values expanded into building the trust would become an outstanding point to undertake in helping the achievement towards the organisation goal with the technology inquiry, processes consolidation, and people encouragement.

In addition, the core values of digital ethics' committed comprehension are strategically organised with having a sufficient work to maintain the human life practice foundation [18]. As a result, the contribution of empowering the trust on both information inquiry and dissemination process is aimed at gaining an easy way to keep in mind of the digital transparency. The sufficient skill and practice of digital ethics come from an overarching set of values in addressing the technology advancement rules including the entire process, the users' engagement and organisation's approach to its use of technologies [19]. As the way in enabling the core guideline of driving the organization and business operations and strategy, an active engagement in considering the principles has to go through assessing the knowledge comprehension on a proactively perceived technology use. With this regard, the extent of technological practices and skills in a way fitted to the core values and fundamental purpose is considerably strengthened in underlying the digital ethics' regulation [20]. The process of shifting the appropriate compliance has to be fitted into the digital ethics' core values in ensuring all connected lines in a way to build the strategic approach of cyber security. The important and significant contribution has to place into maintaining the transparency assurance followed with privacy concern in all online practices [2, 5]. It is important to take a value in guiding the approach of ethical technology concerned in complementing the digital ethics' approach together with serving a substantial extension on the digital atmosphere. The inquiry process provided during dealing with the customers, for instance, should go through a strategic approach in the way of expanding the personal information [8]. As a result, it is necessary to have a sufficient detail on tracking an online behaviour in the attempts on taking an account into digital skills and practices stability.

14.2.2 Digital Ethics for Information Trust Quality

The essence of digital ethics in enabling the organisation what to do within an online platform is required to have a common line on taking the behavioural attitude. With substantial compliance on information accuracy, the digital ethics' committed agreement is supposed to enhance the underlying performance amongst the users on online practices [21]. The stability encouragement on advancing the information accuracy together with the wide range of security and privacy is transmitted to offer the digital service in advocating the human life interaction online. In the attempts to integrate the privacy concern, for instance, the extent of information accuracy could be empowered in advocating the challenge of misconception and misuse of data amongst the companies [37]. With this regard, the consistency of encountering the offer with the mediated service, the digital ethics' information accuracy might be generated in the revenue process in enabling the digital users to have an early knowledge comprehension concerning the digital society issues and solving strategy arrangement [60, 68]. Both sufficient comprehension and practical skills on conceptualising ethical values in underlying what to do in line with an online practice should be taken into consideration in a particular way [35]. The main occupation of providing the digital ethics' further detail is widely organised to commit within the digital online practice. It is sure to have a complete engagement with an online environment in the support of complementing both ethical and professional balance [28]. The further inquiry of looking into detail about whether they will be engaged in the support of digital information could be incorporated with the online platform.

In addition, the strategic effort to enhance the digital ethics' committed practice played a significant role in continuing the information accuracy achievement. As a result, the digital circumstance is required to have a substantial side-step on contributing the trust in the sense that the legal use should be applied [31]. For instance, the continuing process of determining the government agencies is supposed to contribute acquiring the detailed information to consistently coninue in maintaining the security concern [43, 47]. The strategic commitment on practising the ethical discourse reflected into digital environment should bring along with advancing the continued stability to instructing the logical value on online practice for the various purposes [39]. In particular, the careful engagement with influencing the users of adapting and using digital platforms is widely transmitted into digital ethics' commitment in underlying the online practices. Attempts to advance both personal and social capacity in addressing the digital ethics are required to engage with the capability on comprehending and practising what to do in line with the digital practices [41]. Addressing the digital ethics reflected into both skills and practices amidst the digital circumstances requires comprehending the critical worldview in enabling the digital users to achieve an actively equitable outcome. With this regard, the digital circumstances should be initiated in determining the merit of performing the clear instruction provided explicitly within the digital ethics committed management.

In line with having a clear point of expanding the strategic transmission of ethical engagement amidst the digital environment, the large scale of digital inclusion stability in providing the continued service with the data policy should be prominently displayed in a proper manner. As a result, it is sure to make an easy way to comprehend advocates contending the principle to gain a particular attention amongst the digital users by addressing the necessary act to carry out [42]. The traction on widely implementing the digital ethics' information transparency in the organisation sector should be committed in providing the consistency and relevance during the online practice. With this regard, the consequence in applying for the digital ethics reflected into the ethical discourse commitment is clearly showed through the social networking site (SNS) platform arrangement scenario [43, 47]. As such, the strategic online practice is required to make a stability on following the digital ethics' core values. An active engagement should be paid with a particular attention on making stable of the moral obligations reflected from the digital ethics framework and commitment. In this view, the careful awareness engaged with imposing the digital users is pointed out organising the points deserving the careful thought on digital users' parts. In the attempts to assist the digital ethics' comprehension on obtaining the information accuracy, the continued development of embedding the conceptual framework detail to actual practice should be taken into consideration in particular [44]. In consequence, navigating the digital ethics' strategic assurance in taking into an actual practice in giving the clear information.

14.2.3 Digital Ethics as Platform for Gender Violence Anticipation

Referring to point out looking into detail about the way on how to manage oneself ethically, the main consequence of creating the environment with digital ethics' partnership is linked professionally and clinically engaged into building the online and digital platform [45]. The ethical sound manner with having the digital environment is strategised in an appropriate basis. With this regard, the extent of reconstructing both accessibility and adaptability of online practice comes from the near outlook on issues. As a result, the emerging point of increasing the digital inclusion in the partnership environment should be governed in improving the online users' quality [48]. The digital ethics' partnership environment to enhance the accessibility on online practice is supposed to contribute in improving the quality of people's everyday lives. Attempts to commit with the ethical principles through achieving the transparency with the accountability are required in building the instrumental tools aligned to adjust both services and products [49]. On this view, the essential value on building the strategic commitment on digital trust is reflected to build the online practice with the confidence. The strategic value on expanding the digital environment with the ethical discourse engagement needs to focus on getting the solving skills towards gender violence anticipation [51]. Through stabilising the digital ethics' partnership environment, the main consequence and occupation are enhanced with building the response on the ethical engagement.

In addition, the strategy of enhancing digital ethics' partnership has to be given a full attention in ensuring the online practice within the proper digital platform. Moreover, attempts to obtain the performance in stabilising the work commitment with the responsible awareness are flourished in enhancing the innovation arrangement [52]. In particular, the strategic enhancement to have the digital ethics' commitment should bring along with creating the basis of human life relations and activities within an online platform. With this regard, it is necessary to point out advancing the extent of obtaining the optimal model of fairly complemented skills within the human relations. On this view, the ethical discourse engaged to build the foundation amongst digital society's moral consciousness is required to share the online principle to achieve mutual comprehension with social harmony [33]. In the attempts to have disseminated the rules on obtaining the mutual understanding with the harmonious consequence, the stability of engaging the ethical principles is needed to involve into constructing way to interact and live in a digital world [29]. The style on building the online relationship in underlying the digital online practice is enhanced with integrating the platform variety such as messengers, chats, social media, and others in order to transmit the news and information [53]. In order to expand the way of digital communication, it is necessary to have a look into seeing and sharing the data amongst the digital user, in which they are required to simultaneously give the useful information among them. In this view, the way on how the evolvement of such principles in an online platform is globalised world in enabling the process of information transparency identification in the digital age is supposed to clearly demonstrate the digital users' engagement with an active participation [54]. With bringing digital skills and practices through ethical discourse, the effort to embed the inner capacity and stability of digital life circumstance should be taken into consideration in fulfilling the need for sharing the information and data accuracy.

In line with advancing the digital ethics' partnership commitment for information transparency, the strategic determination of enabling the critical use and adaptability could be developed further in producing the rules on online practice. With this regard, creating the condition of partnership commitment on digital ethics refers to apply for developing the ethicalness in featuring the digital users' association with their peers in online [55]. The stability in having a sufficient engagement of digital ethics to give the potentials of technological benefit and solution should do with bringing the ethical principles adapted in online practice. As a result, the strategic commitment on creating the clear information played a significant role in achieving the transparency growth with the digital ethics over the conventional mode. On this view, the necessary act to adopt in developing the ethical principles should be adopted with addressing the guideline standard [56]. In further, the stability of organising the online actual practice should do with building the strategic comprehension on driving the pathway of digital ethics' core values in achieving the information transparency [22]. An active engagement on navigating the critical issues of individual and social concern with the digital ethics is supposed to contribute in dealing with the appropriate mediation on underlying the digital practice [57]. The attempts to pursue the mindfulness on reflecting the digital ethics are required to the determination of creating the information transparency.

14.2.4 Sustaining Digital Ethics Through Information Quality for Gender Violence Anticipation

Attempts to continue expanding the digital ethics' information quality are required to have a substantial commitment in contributing the violence prevention, mainly gender-based initiative. The number of widely strategic approaches on providing the clear details of information has to bring the potentials of addressing data information and digital ethics [61]. Moreover, the critical look on having a clear point of data ethics could be viewed into several phases such as data ethics, digital ethics, and also information ethics. With regard to the diverse backgrounds of each country, the role of continuing information quality achievement with digital ethics' arrangement might be varied in relation to the political arrangement and cultural tradition [64]. The effort to establish the data ethics strategy on artificial intelligence for instance would become the detailed example as for reflecting and operating the digital ethics' principles. As a result, continuing process should come up with accumulating information data with addressing the ethical discourse engagement [24]. On this view, the main occupation is to upgrade the ethic's engagement with digital environment space in providing the approvals for the artificial intelligence strategy [65]. The clear declaration of operating the digital ethics' principles refers to the digital values interpreted into the extent of intellectual capacity and cultural tradition and history where attempts to establish the ethical engagement within online practice should be taken into consideration.

In addition, the continuing procedure on bringing the digital ethics arrangement into online practice is potentially initiated to begin with establishing both significant impacts and its challenging boundaries. Moreover, the consistency on adapting the ethical issues within the digital practice and process has to be collaborated with an early creation on performing a significant contribution to the digital design together with encouraging the beneficial and appropriate applications on digital environment space [67]. As a result, the strategic systems in ensuring the digital users to have a safety concern apart from the harmful consequence should do with developing the digital services associated with considering the digital platform arrangement in line with the principle scenario of information transparency and accuracy [7]. In the capacity of finding the proper way on taking a further consideration on an online platform, it is necessary to have sufficient adaptation in making the potential chances to produce the general principles of information accuracy and transparency [68]. With this regard, paying a particular attention should be paid on posing the information content with a clearly justified clarity, where the arrangement of personal and social capacity would be advanced with adopting the digital technology and their use in public services.

Furthermore, the assessment on helping the users and stakeholders on exploring the digital advancement of artificial intelligence (AI) should consider both positive and negative impacts. It is important to have such online tool and assessment in helping to deal with governing the clear ethics and values principle regulation. Such digital ethics' guidelines in building the robots ethics for instance would call to develop ethical policies in focusing on digital adoption and adaptation in order to emphasise the need to have a balance between common good and protecting human dignity [70]. In terms of promoting the digital ethics' strategic arrangement, its beneficial value aims to expand the responsibility with the capacity of formulating the digital data ethics with an appropriate approach [58]. The significances of expanding the initiative on linking into the institutional arrangements together with cultural traditions are placed in emphasising the development of autonomous vehicle as an attempt to present the comprehensive point of view about digital ethics [71]. With this regard, putting a pathway on taking a guide way in governing the public official and society at large requires an approach to have an inquiry process on digital ethics for information transparency. Moreover, the inquiry of dealing with the robotic arrangement for instance should be clearly enhanced with having a legal procedure in stipulating the principles of the information dissemination amidst the digital society [72]. In continuing the critical inquiry and exploration, the strategic rational of approaching both debate and discussion on digital data and technology ethics is undoubtedly expanded with developing and penetrating into all human life aspects.

14.2.5 Rebuilding Digital Ethics on Information Trust for Gender Violence Potentials Prevention

In the attempts to organise digital ethics' information trust, the main occupation aims to advance the initiative to collect data for a wide range of varieties such as service users and customers in having a wide consideration on gender violence potentials. Moreover, it is necessary to have a sufficient comprehension on considering the various activities through data transparency with the trust transmitted in the organisations [73]. Through going further in the way to enhancing the digital ethics, the detailed information trust indicator in achieving an issue of transparency can be made through giving an informed consent for instance in order to use the available data. With this regard, having full comprehension on using and adapting the data with addressing transparency and trust concern should be put with a clearly ethical judgement [74]. In particular, the transparency in bringing an enough supplement on driving the intentions to have the permission to use data should be incorporated with the data security enhancement with a regulatory requirement of compromising the matter of digital ethics. It is important to take note that considering any personal data with its privacy achievement would lead to enhance the respect in ensuring the critical look into appropriate manner could be managed in a proper arrangement [6, 58]. In particular, gaining the potentials of implying for the data integrity and trust should be maintained within the steps which could be taken through expanding the audit procedure for the robust governance regulation on gender violence [75].

In line with promoting the intentional procedure on behavioural commitment on digital environment, it is necessary to point out disseminating an accuracy of data in preventing the misuse potentials by the human society at large. Moreover, the confirmation bias resulted in the attempts to approve the intentional value of belief and practice on digital ethics' information transparency should be enhanced with having a proper solution [4]. The strategic confirmation to include the data transparency is required to go further on existing both belief and expectation on taking a note of information which may be contradicted. With this regard, the way on how to achieve data transparency in the attempts to provide a necessary act on gender violence prevention refers to have a sufficient expansion in expanding the complex systems with the algorithmic components [8]. The extent of having the algorithmic confirmation could be enhanced in creating the situation where a charity is fixed in assisting the certain groups of people. In particular, the arrangement of digital instrument through variety of tools is provided with having to deserve the assistance which lead to manage behavioural ethics within the digital environment space [72]. As a result, attempts to advance the digital ethics culture for the purpose of safety-based digital environment for gender violence anticipation are regulated with the framework guideline to work with the digital team [9]. In order to ensure the digital users in optimizing well in line with the objective, the iniative on monitoring the effort to enable the employees in any organisation should begin with enhancing the sufficient comprehension on building the dimension of digital ethics within any of digital project provided from the program arrangement.

In addition, the approach to apply for the organisation procedure through arranging the code of digital ethics has to be taken into consideration in modifying the ethical guideline with the concern of societal life purpose. The arrangement of modification could be made through expanding the digital ethics commitment in driving on how the concerns on performing digital practice are potentially behaving the particular attention in facing the challenges and controversial situation [10]. As a result, the digital ethics arrangement within the digital organisations should have taken up a clear ethical position in underlying the trust commitment and achievement amongst the users, such as business partners, customers, and suppliers. The variety of ethical aspects in driving the digital ethics arrangement comes with engaging an active performance scale including respect, trustworthiness, fairness, and also caring incorporated within the digital practice [11]. With this regard, the ethical engagement on underlying the workplace is required to have a substantial inclusion on driving the responsibility commitment paid in the online practice arrangement [40, 46]. The major concern on continuing the digital ethics responsibility might be placed as a valuable insight to assist the digital process within the online environment [61]. Moreover, the confirmation on driving the online-based behavioural ethics could be provided with initiating the digital technologies adapted in the organisation sector for instance. As such, the chief information security arrangement should bring along with having the digital team within the organisations in order to ensure the extent of ethical considerations are assessed into the proper manner with an appropriate attention in digital environment space [12].

14.3 Implications and Future Directions

The challenging issues on technology are widely involved along with the case issues such as digital security attack, illegal observation, personal data misuse, misinformation with the algorithmic bias and also transparency crisis [9]. With this regard, the strategic pathway on handling the typical point of having a committed transparency should be incorporated with a digital ethics' committed practice for information trust assurance in line with the human wellbeing sustainability mainly on accuracy and transparency [62, 63]. As such, the distrust is strategically required to get a clear picture about recommending the response towards these challenges breed in the provider, implementer or receiver side in which the significant contribution should be made in organising the digital ethics' committed practice for information trust [10]. The wide range of expanding the trust in information might value to the prevention insight of cyber violence, including gender-based violence. The demand of serious concern about the gender violence has been widely sounded due to the rising number of challenges in the recent years, one of which is on the gender-based consumer trust in varied responses to the commercial enterprises [11]. As the emerging trends of declining phenomena in the failing down amongst the digital users in particular and citizens in general, such cyber challenges would lead to the wary of public sphere. As a result, the strategic requirement to work with all human necessity to conduct for the life purpose should be committed into the practical stability in expanding the digital ethics.

In addition, the necessary point in having the sufficient comprehension of building the digital ethics' committed practices is considered to play a role in stabilising the digital information trust [12]. As such, this chapter aims to examine the strategic approach on empowering digital ethics' committed practice for information trust as an attempt to give a critical insight into valuing the gender violence anticipation. The earlier phase refers to highlight the digital ethics and information accuracy, followed by the digital partnership environment and digital ethics as the driving pathway of managing gender violence anticipation. The main outcome of this procedure might be particularly relevant for the variety of charities in the attempts to build the trust and transparency in their digital content [70]. On this view, it is crucial to have an effective communication about the significance of the digital-based workplace arrangement together with avoiding the readers about the doubting towards the messages veracity. The particular significance would lead to give insight into providing the guideline of digital ethics' information transparency in combatting the gender violence potentials [75]. The digital organisation is arranged with building the digital ethics' approval towards transforming the support and confidence on the digital practice arrangement [14]. In order to achieve this, it is important that organisations pay attention to digital ethics on time in the transformation process and make it an inseparable part of their digital online practice and operation [1, 2, 5, 7, 22–27, 29, 34, 35, 37, 40, 43, 46, 47, 50, 58, 60, 66, 69, 76]. The focus is on their willingness to take responsibility for complying with the digital ethics referring to the extent of moral values arrangement amidst the public sphere regulation [3, 77].

In further, the stability on paying attention with being mindful of expanding the selection on taking benefit and avoiding the harmful impact to others should come up with creating the decision on the post arrangement [59]. In empowering the basic principles on digital online practice, the application rules in further determining the actual point of decision-making process are required to enhance the global reaching enhancement with a corporate digital ethics. The strategic recommendation on focusing on assisting the digital online practice and the public service arrangement has to bring along with governing the processes and officials in providing the guiding principles to ensure ethical use adaptation and adoption with developing a comprehensive assessment on online practice [72]. As such, the clear stability on reaching the key principles of continuing digital ethics for information transparency in assisting the gender violence anticipation should be appropriately governed in enabling the ethical engagement on businesses, government and individuals to openly proceed the serving purpose with an awareness and knowledge-sharing mechanism [12]. Through ensuring the data transparency from the alteration or deletion, the need to proceed in a proper way should be taken into consideration in undertaking the beneficial value and maintaining from the harmful potentials to provide the properly solving initiative arrangement [30, 32, 78]. The attempts to build the digital ethics' information trust in preventing gender violence are assessed in resulting the ethical behaviour within the online environment basis [74]. In undertaking the analytical performance of data, carrying out its potentials to give a contribution into providing the gender violence anticipation could be made with considering the ethical examination in guiding the behaviour.

14.4 Conclusion

Since the number of violence in the digital age has been gradually emerged with the variety of forms, approaches and motives, such condition needs to have a careful engagement in responding properly and effectively disseminating the positive feedback in giving insights mainly into the gender violence prevention. Such prevention needs to pay attention on reconstructing the application strategy. The main concern of this study was on reframing the digital ethics' main features in underlying the attempts for information trust in enabling the users to have interaction support amidst the digital circumstance. This chapter aims to examine the demanding needs of digital ethics elaborated as a strategic foundation to expand safety concern amongst digital community and partnership for gender violence anticipation. The significant alignment of digital ethics in expanding the digital partnership skills has a core value to expand the digital information trust in the order for gender violence anticipation. Those are continuing digital ethics' information quality for gender violence anticipation, rebuilding digital ethics' information trust to prevent gender violence, and promoting digital ethics' grant on digital safety environment assurance for gender violence anticipation. The current study aims to give the value on building the digital technology enhancement to advance the digital society in driving their purposes, motives, and business orientation with the professional and ethical pathway.

References

1. Abdul Aziz, N. A., Mohd Razali, F., & Saari, C. Z. (2022). Penggunaaan Media Sosial dari Perspektif Psiko Spiritual Islam. *Firdaus Journal, 2*(1), 65–75. https://doi.org/10.37134/firdaus.vol2.1.6.2022
2. Ab Rahim, N. M. Z., Saari, Z., Mohamad, A. M., Rashid, M. H., & Mohamad Norzilan, N. I. (2022). Konsep Ulul Albab dalam Al-Quran dan hubungannya dengan pembelajaran kursus Sains, Teknologi dan Manusia di UTM Kuala Lumpur. *Firdaus Journal, 2*(2), 72–78. https://doi.org/10.37134/firdaus.vol2.2.7.2022

3. Adil, M., & Huda, M. (2023). Understanding Responses to Worship Regulations in the Pandemic Era: Text Data Mining Analysis in the Indonesian Context. *Religions, 14*(4), 549.
4. Almås, A. G., Bueie, A. A., & Aagaard, T. (2021). From digital competence to professional digital competence: Student teachers' experiences of and reflections on how teacher education prepares them for working life. *Nordic Journal of Comparative and International Education (NJCIE), 5*(4), 70–85.
5. Alwi, A. S. Q., & Ibrahim, R. (2022). Isu terhadap Penggunaan Teknologi Media Digital dalam kalangan guru pelatih jurusan Pendidikan Khas. *Firdaus Journal, 2*(2), 88–93. https://doi.org/10.37134/firdaus.vol2.2.9.2022
6. Anshari, M., Almunawar, M. N., Shahrill, M., Wicaksono, D. K., & Huda, M. (2017). Smartphones usage in the classrooms: Learning aid or interference? *Education and Information Technologies, 22*(6), 3063–3079.
7. Arifin, S., Huda, M., & Mufida, N. H. (2023). Akhlak Karimah Value-Based Integrative Learning Model in Madrasah. *Jurnal Pendidikan Islam. 9*(1), 41–54.
8. Bartolomé, J., Garaizar, P., & Larrucea, X. (2021). A pragmatic approach for evaluating and accrediting digital competence of digital profiles: A case study of entrepreneurs and remote workers. *Technology, Knowledge and Learning, 27*, 1–36.
9. Beardsley, M., Albó, L., Aragón, P., & Hernández-Leo, D. (2021). Emergency education effects on teacher abilities and motivation to use digital technologies. *British Journal of Educational Technology, 52*(4), 1455–1477.
10. Behnamnia, N., Kamsin, A., Ismail, M. A. B., & Hayati, A. (2020). The effective components of creativity in digital game-based learning among young children: A case study. *Children and Youth Services Review, 116*, 105227.
11. Cabezas-González, M., Casillas-Martín, S., & García-Peñalvo, F. J. (2021). The digital competence of pre-service educators: The influence of personal variables. *Sustainability, 13*(4), 2318.
12. Cattaneo, A. A., Antonietti, C., & Rauseo, M. (2022). How digitalised are vocational teachers? Assessing digital competence in vocational education and looking at its underlying factors. *Computers & Education, 176*, 104358.
13. Colás-Bravo, P. C. B., Conde-Jiménez, J. C. J., Reyes-de, S. R. D. C., Colás-Bravo-Bravo, P., Conde-Jiménez, J., & Reyes-de-Cózar, S. (2019). The development of the digital teaching competence from a sociocultural approach. *Comunicar: Media Education Research Journal, 27*(2). https://doi.org/10.3916/C61-2019-02
14. Colás-Bravo, P., Conde-Jiménez, J., & Reyes-de-Cózar, S. (2021). Sustainability and digital teaching competence in higher education. *Sustainability, 13*(22), 12354.
15. Darazha, I., Lyazzat, R., Ulzharkyn, A., Saira, Z., & Manat, Z. (2021, March). Digital competence of a teacher in a pandemic. In *2021 9th international conference on information and education technology (ICIET)* (pp. 324–328). IEEE.
16. Galindo-Domínguez, H., & Bezanilla, M. J. (2021). Promoting time management and self-efficacy through digital competence in university students: A mediational model. *Contemporary Educational Technology, 13*(2), ep294.
17. Galindo-Domínguez, H., & Bezanilla, M. J. (2021). Digital competence in the training of pre-service teachers: Perceptions of students in the degrees of early childhood education and primary education. *Journal of Digital Learning in Teacher Education, 37*(4), 262–278.
18. Gordillo, A., Barra, E., Garaizar, P., & López-Pernas, S. (2021). Use of a simulated social network as an educational tool to enhance teacher digital competence. *IEEE Revista Iberoamericana de Tecnologias del Aprendizaje, 16*(1), 107–114.
19. Guillén-Gámez, F. D., Mayorga-Fernández, M., Bravo-Agapito, J., & Escribano-Ortiz, D. (2021). Analysis of teachers' pedagogical digital competence: Identification of factors predicting their acquisition. *Technology, Knowledge and Learning, 26*(3), 481–498.
20. Guillén-Gámez, F., Cabero-Almenara, J., Llorente-Cejudo, C., & Palacios-Rodríguez, A. (2021). Differential analysis of the years of experience of higher education teachers, their digital competence and use of digital resources: Comparative research methods. *Technology, Knowledge and Learning, 27*, 1–21.

21. Hanafi, H. F., Wahab, M. H. A., Selamat, A. Z., Masnan, A. H., & Huda, M. (2020, November). A systematic review of augmented reality in multimedia learning outcomes in education. In *International conference on intelligent human computer interaction* (pp. 63–72). Springer.
22. Huda, M. (2019). Empowering application strategy in the technology adoption: insights from professional and ethical engagement. *Journal of Science and Technology Policy Management, 10*(1), 172–192. https://doi.org/10.1108/JSTPM-09-2017-0044
23. Huda, M. (2022). Empowering professional and ethical balance in digital record management. *Organizational Cybersecurity Journal: Practice Process and People, 2*(1), 60–73. https://doi.org/10.1108/OCJ-06-2021-0016
24. Huda, M. (2022). Towards an adaptive ethics on social networking sites (SNS): A critical reflection. *Journal of Information, Communication and Ethics in Society, 20*(2), 273–290. https://doi.org/10.1108/JICES-05-2021-0046
25. Huda, M. (2023). Trust as a key element for quality communication and information management: insights into developing safe cyber-organisational sustainability. *International Journal of Organizational Analysis*. https://doi.org/10.1108/IJOA-12-2022-3532
26. Huda, M. (2023). Towards digital access during pandemic age: better learning service or adaptation struggling? *Foresight, 25*(1), 82–107. https://doi.org/10.1108/FS-09-2021-0184
27. Huda, M. (2023). Digital marketplace for tourism resilience in the pandemic age: voices from budget hotel customers. *International Journal of Organizational Analysis, 31*(1), 149–167. https://doi.org/10.1108/IJOA-10-2021-2987
28. Huda, M., & Hashim, A. (2022). Towards professional and ethical balance: insights into application strategy on media literacy education. *Kybernetes, 51*(3), 1280–1300.
29. Huda, M., & Salem, S. (2022). Maintaining Quality Family Time for Children's Social Intelligence: Public Educators' Beliefs and Practices in the Pandemic Age. In *Children, Youth and Time* (Vol. 30, pp. 69–90). Emerald Publishing Limited.
30. Huda, M., & Salem, S. (2022). Understanding Human Behavior Development with Spirituality: Critical Insights into Moral Flourishing. *Ulumuna, 26*(2), 238–268.
31. Huda, M., & Teh, K. S. M. (2018). Empowering professional and ethical competence on reflective teaching practice in digital era. In *Mentorship strategies in teacher education* (pp. 136–152). IGI Global.
32. Huda, M., Arifin, S., Ali, A. H., Selamat, A. Z., Ibrahim, M. H., Hashim, A., Isa, N .K. M., & Ab Rahman, Z. (2023). Enhancing Digital Leadership Direction: Insight into Empowering Gender Violence Prevention. Deepanjali Mishra et al. (Eds): *Communication Technology and Gender Violence*. Springer.
33. Huda, M., Muhamad, N. H. N., Isyanto, P., Kawangit, R. M., Marni, N., Mohamed, A. K., & Safar, A. J. (2020). Building harmony in diverse society: Insights from practical wisdom. *International Journal of Ethics and Systems., 36*(2), 149–165. https://doi.org/10.1108/IJOES-11-2017-0208
34. Huda, M., Ramli, A. F., Ritonga, M., Rahim, M. M. A., Borham, A. H., Hashim, A., Husain, H., & Isa, N. K. M. (2023a). From Digital Ethics to Digital Partnership Skills: Driving a Safety Strategy to Expand the Digital Community?. Digital Transformation for Business and Society: Contemporary Issues and Applications in Asia. UK: Taylor & Francis.
35. Huda, M., Ramli, A. F., Ritonga, M., Rahim, M. M. A., Borham, A. H., Hashim, A., Hanafi, H. F., & Isa, N. K. M. (2023b). Towards Digital Servant Leadership for Organisational Stability: Driving Processes in the Pandemic Age?. Digital Transformation for Business and Society: Contemporary Issues and Applications in Asia. UK: Taylor & Francis.
36. Huda, M., Rosman, A. S., Mohamed, A. K., & Marni, N. (2021). Empowering adaptive learning technology: Practical insights from social network site (SNS). In R. Silhavy, P. Silhavy, & Z. Prokopova (Eds.), *Software engineering application in informatics. CoMeSySo 2021* (Lecture notes in networks and systems) (Vol. 232). Springer. https://doi.org/10.1007/978-3-030-90318-3_34
37. Huda, M., Sutopo, L., Liberty, Febrianto, & Mustafa, M. C. (2022). Digital information transparency for cyber security: critical points in social media trends. In Advances in Information

and Communication: Proceedings of the 2022 Future of Information and Communication Conference (FICC), Volume 2 (pp. 814–831). Cham: Springer International Publishing.
38. Iglesias-Rodríguez, A., Hernández-Martín, A., Martín-González, Y., & Herráez-Corredera, P. (2021). Design, validation and implementation of a questionnaire to assess teenagers' digital competence in the area of communication in digital environments. *Sustainability, 13*(12), 6733.
39. Isoda, M., Estrella, S., Zakaryan, D., Baldin, Y., Olfos, R., & Araya, R. (2021). Digital competence of a teacher involved in the implementation of a cross-border lesson for classrooms in Brazil and Chile. *International Journal for Lesson & Learning Studies*. https://doi.org/10.1108/IJLLS-05-2021-0045
40. Jailani, M., Hafidh, N., & Huda, M. (2023). The Influence of Upin And Ipin Animation on Children's Religious Life. *Penamas, 36*(1), 74–90.
41. Kalimullina, O., Tarman, B., & Stepanova, I. (2021). Education in the context of digitalization and culture: Evolution of the teacher's role, pre-pandemic overview. *Journal of Ethnic and Cultural Studies, 8*(1), 226–238.
42. Khalili, H. M., Rosman, A. S., Mohamed, A. K., & Marni, N. (2021). Digital learning enhancement through social network site (SNS). In R. Silhavy, P. Silhavy, & Z. Prokopova (Eds.), *Software engineering application in informatics. CoMeSySo 2021* (Lecture notes in networks and systems) (Vol. 232). Springer. https://doi.org/10.1007/978-3-030-90318-3_35
43. Khalili, Huda, M., Rosman, A. S., Mohamed, A. K., & Marni, N. (2021). Digital learning enhancement through social network site (SNS). In *Software Engineering Application in Informatics: Proceedings of 5th Computational Methods in Systems and Software* 2021, Vol. 1 (pp. 421–431). Springer International Publishing.
44. Kilic, F., & Karakuş, I. (2021). New features of learners in education: Digital awareness, digital competence, and digital fluency. In *Improving scientific communication for lifelong learners* (pp. 113–132). IGI Global.
45. Lohr, A., Stadler, M., Schultz-Pernice, F., Chernikova, O., Sailer, M., Fischer, F., & Sailer, M. (2021). On powerpointers, clickerers, and digital pros: Investigating the initiation of digital learning activities by teachers in higher education. *Computers in Human Behavior, 119*, 106715.
46. Maimun, M., Huda, M., & Muhammad, M. Strategies for Building A Competitive and Excellent Islamic Educational Institution. *POTENSIA: Jurnal Kependidikan Islam, 8*(2), 137–153.
47. Maseleno, A., Huda, M., Jasmi, K. A., Basiron, B., Mustari, I., Don, A. G., & bin Ahmad, R. (2019). Hau-Kashyap approach for student's level of expertise. *Egyptian Informatics Journal, 20*(1), 27–32.
48. Mehrvarz, M., Heidari, E., Farrokhnia, M., & Noroozi, O. (2021). The mediating role of digital informal learning in the relationship between students' digital competence and their academic performance. *Computers & Education, 167*, 104184.
49. Mishra, D., & Sain, M. (2021). Role of digital education to curb gender violence. In *Strategies for e-service, e-governance, and cybersecurity: Challenges and solutions for efficiency and sustainability* (Vol. 33). Routledge.
50. Mulyadi, D., Huda, M., & Gusmian, I. (2022). Smart Learning Environment (SLE) in the Fourth Industrial Revolution (IR 4.0). *International Journal of Asian Business and Information Management, 13*(2), 1–23. https://doi.org/10.4018/IJABIM.20220801, https://doi.org/10.4018/IJABIM.287589
51. Nasr, H. A. (2020). Competences in digital online media literacy: Towards convergence with emergency remote EFL learning. *International Journal of Media and Information Literacy, 5*(2), 164–175.
52. Olesika, A., Lama, G., & Rubene, Z. (2021). Conceptualization of digital competence: Perspectives from higher education. *International Journal of Smart Education and Urban Society (IJSEUS), 12*(2), 46–59.
53. Örtegren, A. (2022). Digital citizenship and professional digital competence—Swedish subject teacher education in a postdigital era. *Postdigital Science and Education, 4*, 1–27.
54. Passey, D., Bottino, R., Lewin, C., & Sanchez, E. (2019). *Empowering learners for life in the digital age*. Springer.

55. Pérez-Calderón, E., Prieto-Ballester, J. M., & Miguel-Barrado, V. (2021). Analysis of digital competence for Spanish teachers at pre-university educational key stages during COVID-19. *International Journal of Environmental Research and Public Health, 18*(15), 8093.
56. Pérez-Navío, E., Ocaña-Moral, M. T., & Martínez-Serrano, M. D. C. (2021). University graduate students and digital competence: Are future secondary school teachers digitally competent? *Sustainability, 13*(15), 8519.
57. Petrushenko, Y., Onopriienko, K., Onopriienko, I., & Onopriienko, V. (2021, September). Digital learning for adults in the context of education market development. In *2021 11th international conference on advanced computer information technologies (ACIT)* (pp. 465–468). IEEE.
58. Qadafi, M., Ulfah, M., Huda, M., & Agustiningsih, N. Fostering Independent Learning in Early Childhood: A Case Study on Montessori Pedagogy at PAUD Montessori Futura Indonesia. *Golden Age: Jurnal Ilmiah Tumbuh Kembang Anak Usia Dini, 8*(3), 109–120.
59. Quaicoe, J. S., & Pata, K. (2020). Teachers' digital literacy and digital activity as digital divide components among basic schools in Ghana. *Education and Information Technologies, 25*(5), 4077–4095.
60. Rachman, A., Oktoviani, I., & Manurung, P. (2023). Digitization of Zakat and Charity BAZNAS Tangerang City through Crowdfunding Platform tangerangsedekah. id. *Firdaus Journal, 3*(1), 96–106. https://doi.org/10.37134/firdaus.vol3.1.9.2023
61. Rahayu, N. W., & Haningsih, S. (2021). Digital parenting competence of mother as informal educator is not inline with internet access. *International Journal of Child-Computer Interaction, 29*, 100291.
62. Ramli, A. F., Ramli, N., Huda, M., Asy-Syaimaa'Hussain, L. K., & Ramli, N. I. A. (2022). Critical Investigation on the Pandemic from the Islamic Perspective. *Afkar: Jurnal Akidah & Pemikiran Islam*, 99–140.
63. Ramle, M. R., & Huda, M. (2022). Between Text and Context: Understanding Ḥadīth through Asbab al Wurud. *Religions, 13*(2), 92.
64. Reisoğlu, İ. (2021). How does digital competence training affect teachers' professional development and activities? *Technology, Knowledge and Learning, 27*, 1–28.
65. Robles Moral, F. J., & Fernández Díaz, M. (2021). Future primary school teachers' digital competence in teaching science through the use of social media. *Sustainability, 13*(5), 2816.
66. Ropei, A., Huda, M., Alijaya, A., Fadhil, F., & Zulfa, F. (2023). Managing 'Baligh'in Four Muslim Countries: Egypt, Tunisia, Pakistan, and Indonesia on the Minimum Age for Marriage. *Al-Ahwal: Jurnal Hukum Keluarga Islam, 16*(1), 112–140.
67. Sailer, M., Murböck, J., & Fischer, F. (2021). Digital learning in schools: What does it take beyond digital technology? *Teaching and Teacher Education, 103*, 103346.
68. Sailer, M., Stadler, M., Schultz-Pernice, F., Franke, U., Schöffmann, C., Paniotova, V., et al. (2021). Technology-related teaching skills and attitudes: Validation of a scenario-based self-assessment instrument for teachers. *Computers in Human Behavior, 115*, 106625.
69. Sari, D. C., Ali, A. H., Harun, M., Batre, N. M., Hanafi, M. S., Jaludin, Z. Y., ... & Kuswandi, I. (2023). Transformation of Artificial intelligence in Islamic Edu with Ulul Albab Value (Global Challenge Perespective). *Firdaus Journal, 3*(1), 1–9.
70. Shohel, M. M. C., Shams, S., Ashrafuzzaman, M., Alam, A. S., Al Mamun, M. A., & Kabir, M. M. (2022). Emergency remote teaching and learning: Digital competencies and pedagogical transformation in resource-constrained contexts. In *Handbook of research on Asian perspectives of the educational impact of COVID-19* (pp. 175–200). IGI Global.
71. Shonfeld, M., Cotnam-Kappel, M., Judge, M., Ng, C. Y., Ntebutse, J. G., Williamson-Leadley, S., & Yildiz, M. N. (2021). Learning in digital environments: A model for cross-cultural alignment. *Educational Technology Research and Development, 69*(4), 2151–2170.
72. Sillat, L. H., Tammets, K., & Laanpere, M. (2021). Digital competence assessment methods in higher education: A systematic literature review. *Education Sciences, 11*(8), 402.
73. Spante, M., Hashemi, S. S., Lundin, M., & Algers, A. (2018). Digital competence and digital literacy in higher education research: Systematic review of concept use. *Cogent Education, 5*(1), 1519143.

74. Willermark, S. M. J., & Gellerstedt, M. (2022). Digitalization, distance education, virtual classroom, high school, digital competence, COVID-19, ideal-type analysis. *Journal of Educational Computing Research*. https://doi.org/10.1177/07356331211069424
75. Wong, K. M., & Moorhouse, B. L. (2021). Digital competence and online language teaching: Hong Kong language teacher practices in primary and secondary classrooms. *System, 103*, 102653.
76. Yuan, C., Wang, L., & Eagle, J. (2019). Empowering English language learners through digital literacies: Research, complexities, and implications. *Media and Communication, 7*(2), 128–136.
77. Zainuri, A., & Huda, M. (2023). Empowering Cooperative Teamwork for Community Service Sustainability: Insights from Service Learning. *Sustainability, 15*(5), 4551.
78. Zainuri, A., Sukarno, S., & Huda, M. (2022). Understanding Scientific Literacy and Pedagogy Competence: A Critical Insight into Religious Integration Thinking Skills. *Journal of Educational and Social Research, 12*(1), 274–281.

Correction to: Communication Technology and Gender Violence

Deepanjali Mishra, Anh Ngoc Le, and Zachary McDowell

Correction to:
D. Mishra et al. (eds.), *Communication Technology and Gender Violence*, Signals and Communication Technology, https://doi.org/10.1007/978-3-031-45237-6

The original version of this book has been revised. This book was inadvertently published with an error in preface, incorrect affiliation for Editor Zachary McDowell and incorrect affiliation for Chapter 9 authors. These errors have been corrected.

The updated version of this book can be found at
https://doi.org/10.1007/978-3-031-45237-6
https://doi.org/10.1007/978-3-031-45237-6_9

D. Mishra et al. (eds.), *Communication Technology and Gender Violence*, Signals and Communication Technology, https://doi.org/10.1007/978-3-031-45237-6_15

Index

D. Mishra et al. (eds.), *Communication Technology and Gender Violence*, Signals and Communication Technology, https://doi.org/10.1007/978-3-031-45237-6

GPSR Compliance

The European Union's (EU) General Product Safety Regulation (GPSR) is a set of rules that requires consumer products to be safe and our obligations to ensure this.

If you have any concerns about our products, you can contact us on ProductSafety@springernature.com

In case Publisher is established outside the EU, the EU authorized representative is:

Springer Nature Customer Service Center GmbH
Europaplatz 3
69115 Heidelberg, Germany

Batch number: 10755227

Printed by Printforce, the Netherlands